# Blockchain and Software-Defined Networking

This book covers the security challenges of software-defined networks and proposes innovative solutions, such as the decentralization of security services and integration with blockchain technology, to enhance software-defined network security. It includes multiple case studies demonstrating the effectiveness of decentralized software-defined network security with three-layered architectures including all application programming interfaces.

This book:

- Offers in-depth analysis of software-defined network protocols including OpenFlow, Border Gateway Protocol-Link State, and Network Configuration Protocol (NETCONF).
- Provides practical guidelines for implementing security measures within software-defined network environments, with a focus on technical aspects like firewall rules, intrusion detection, and access control policies.
- Explores how blockchain can be integrated into software-defined networks to enhance security, providing step-by-step technical guidance for implementation.
- Discusses supply chain transparency with integration of blockchain and software-defined networks.
- Explains blockchain-enhanced network monitoring and control, and implementation of blockchain-secured network communication.

It is primarily written for senior undergraduates, graduate students, and academic researchers in diverse fields, including electrical engineering, electronics and communications engineering, computer science and engineering, and information technology.

**Intelligent Data-Driven Systems and Artificial Intelligence**
*Series Editor: Harish Garg*

**Cognitive Machine Intelligence**
*Applications, Challenges, and Related Technologies*
Inam Ullah Khan, Salma El Hajjami, Mariya Ouaissa, Salwa Belqziz, Tarandeep Kaur Bhatia

*Artificial Intelligence and Internet of Things Based Augmented Trends for Data Driven Systems*
Anshu Singla, Sarvesh Tanwar, Pao-Ann Hsiung

*Modelling of Virtual Worlds Using the Internet of Things*
Edited by Simar Preet Singh and Arun Solanki

*Data-Driven Technologies and Artificial Intelligence in Supply Chain Tools and Techniques*
Mahesh Chand, Vineet Jain, Puneeta Ajmera

*Data-Driven Decision Making and Soft Computing Applications and Advances*
D. Nagarajan, Dragan Pamucar, Zarife Zararsız

*Artificial Intelligence Powered Sustainable Development of Business 4.0*
Edited by Hemant Kumar Saini, Sita Rani, Vikram Puri, Noor Zaman Jhanjhi

*Artificial Intelligence and Fuzzy Logic for Next Generation Intelligent Transportation Systems*
Mariya Ouaissa, Mariyam Ouaissa, Najib Ben Aoun, Anass Cherrafi, Rahul Kar

*Blockchain and Software-Defined Networking: Opportunities, Challenges, and Future Trends*
Sarvesh Tanwar, Sumit Badotra, Harish Garg

For more information about this series, please visit: www.routledge.com/Intelligent-Data-Driven-Systems-and-Artificial-Intelligence/book-series/CRCIDDSAAI

# Blockchain and Software-Defined Networking

## Opportunities, Challenges, and Future Trends

Edited by
Sarvesh Tanwar, Sumit Badotra, and
Harish Garg

CRC Press
Taylor & Francis Group
Boca Raton London New York

CRC Press is an imprint of the
Taylor & Francis Group, an **informa** business

Designed cover image: Add credit line: Shutterstock

First edition published 2026
by CRC Press
2385 NW Executive Center Drive, Suite 320, Boca Raton FL 33431

and by CRC Press
4 Park Square, Milton Park, Abingdon, Oxon, OX14 4RN

*CRC Press is an imprint of Taylor & Francis Group, LLC*



ISBN: 9781032824369 (hbk)
ISBN: 9781032824383 (pbk)
ISBN: 9781003504559 (ebk)

DOI: 10.1201/9781003504559

Typeset in Sabon
by codeMantra

## Contents

# Preface

In an era of unprecedented technological advancements, the worlds of blockchain and Software-Defined Networking (SDN) have emerged as transformative forces, reshaping the way we conceptualize and implement network architectures. This book, *Blockchain Enabled Software Defined Networking: Recent Trends and Development*, delves into the convergence of these two dynamic domains, exploring their synergies, implications, and potential to revolutionize our approach to secure, efficient, and resilient networking. In today's rapidly evolving landscape, the demand for secure, scalable, and adaptable networks has never been more pressing. As networks expand exponentially and cyber threats grow in sophistication, the need for innovative solutions has given rise to the fusion of blockchain and SDN. This book, aimed at technology enthusiasts, researchers, professionals, and students alike, aims to shed light on this fusion's intricacies and opportunities.

The journey within these pages traverses the foundational concepts of both blockchain and SDN, unraveling their inner workings and highlighting their strengths and limitations. We explore the challenges inherent in traditional networking models, where centralized control can lead to bottlenecks, vulnerabilities, and single points of failure. Enter blockchain, a decentralized ledger technology, offering a path to enhanced security, transparency, and trust across network ecosystems.

The heart of this book lies in the exploration of how blockchain can fortify SDN security, decentralizing control and ushering in a new era of resilient network management. We delve into the architectural intricacies of integrating blockchain and SDN, examining smart contracts as enablers of automation and trust. Real-world use cases showcase the practical applications of blockchain-secured network management and supply chain tracking, illuminating the transformative potential of this fusion.

We also acknowledge the challenges and considerations that arise. From scalability concerns to ethical implications, we navigate the complexities that accompany innovation. This book aims to provide a comprehensive perspective, discussing future trends, potential impacts on industry landscapes, and legal frameworks that align with this technological convergence.

As editors, we have embarked on this journey with a passion for exploring the uncharted territories where blockchain and SDN intersect. We hope this book sparks curiosity, ignites discussions, and serves as a compass guiding readers through the evolving landscapes of secure networking and technological advancement. We invite you to join us in this exploration, as we navigate the uncharted waters of blockchain-enabled SDN, uncovering insights, challenges, and opportunities that shape the future of networking and security.

# About the editors

**Sarvesh Tanwar** is Professor at Amity Institute of Information Technology (AIIT), Amity University, Noida. She is the head of AUN Blockchain & Data Security Research Lab. She has completed her M.Tech. (CSE) degree from MMU, Mullana, and Ph.D. in (CSE) from Mody University, Laxmangarh (Raj.). She has more than 15 years of teaching and research experience. Her area of research includes Public Key Infrastructure, Cryptography, Blockchain, and Cyber Security. She has published more than 100 research papers in International Journals (SCI/Scopus/WoS and Peer-Review) and Conferences. She is currently guiding six Ph.D. scholars and has supervised five Ph.D. scholars and five M.Tech. research scholars. She has filed 22 patents (8 granted) and 2 copyrights in the relevant field. She is a senior member of IEEE, Life Member, Cryptology Research Society of India (CRSI), Indian Institute of Statistics, Kolkata, India, and member of International Association of Computer Science and Information Technology (IACSIT), Singapore. She is also a member of the Editorial Board of the *International Journal of Sensors, Wireless Communications and Control, Bentham Science*.

**Sumit Badotra** is currently serving as a postdoctoral research fellow at the Department of Electrical and Computer Engineering, National University of Singapore (NUS), Singapore. He received his Ph.D. in Computer Science and Engineering from Chitkara University, India, under a DST-Government of India-funded project, where his research focused on Software-Defined Networking security and anomaly detection frameworks. He has also held academic positions at various universities, contributing to both teaching and funded research activities. Dr. Badotra has authored and coauthored several research papers published in high-impact international journals and conferences. His research interests include Software-Defined Networking security, network anomaly detection, distributed architectures, machine learning, and cloud computing. He has also worked on real-time frameworks for DDoS mitigation, adaptive ML-based security solutions, and hybrid privacy-preserving architectures. He actively contributes as a reviewer and editorial board member for various international journals and conferences in the field of computer networks and cybersecurity. Dr. Badotra holds multiple patents and has received best paper and presentation awards for his contributions to network security research.

**Dr. Harish Garg** working as an Associate Professor at Thapar Institute of Engineering & Technology, Patiala, Punjab, India. He is ranked in the World's Top 2% Scientist List and **India Rank #1 & World Rank #163** published by Stanford University in the consecutive five years 2020-2024. He received the *Most Outstanding Researcher award* in the field of Mathematics from Careers 360 Academy. He is also the recipient of the International Obada-Prize 2022 – Young Distinguished Researchers. He is also the recipient of the Top-Cited paper by an India-based author (2015 – 2019) from Elsevier Publisher. He also serves as an advisory board member of the Universal Scientific Education and Research Network (USERN). He is the Research Fellow of INTI International University, Malaysia.

**Dr. Garg's** research interests include Computational Intelligence, Multi-criteria decision making, Evolutionary algorithms, Reliability analysis, Expert systems and decision support systems, Computing with words and Soft Computing. He has authored more than **530** papers (**over 510 are SCI**) published in refereed International Journals including IEEE Transactions, Elsevier, Springer etc. His Google citations are over **30500 with H-index- 94**. He is one of the leading researchers in the world related to the MCDM and soft computing approaches.

**Dr. Garg** also serves on editorial boards of several leading international journals, this includes the Founding Editor-in-Chief of the *Journal of Computational and Cognitive Engineering*. He is also the Associate Editor of Alexandria Engineering Journal, Journal of Industrial & Management Optimization, Machine Intelligence Research, CAAI Transactions on Intelligence Technology, etc. For more details about him, kindly follow his webpage https://sites.google.com/site/harishg58iitr/home

# List of contributors

**Ahuja Laxmi**
Amity Institute of Information Technology
Amity University
Noida, India

**Badotra Sumit**
Department of Computer and Electrical Engineering
National University of Singapore
Singapore, Singapore

**Bhargava Deepshikha**
ASET
Amity University
Greater Noida, India

**Chidambaram S.**
Department of Information Technology
National Engineering College
Kovilpatti, India

**Das Lipsa**
ASET
Amity University
Greater Noida, India

**Gowthami N.**
Department of Information Technology
National Engineering College
Kovilpatti, India

**Gupta Amardeep**
ASET
Amity University
Greater Noida, India

**Gupta Medini**
Amity Institute of Information Technology
Amity University
Noida, India

**Gurpreet Singh**
School of Computer Science and Engineering
Lovely Professional University (LPU)
Punjab, India

**Kafeel Ahmed Muneer**
Department of Information Technology
C. Abdul Hakeem College of Engineering & Technology
Ranipet, India
and
Accenture India
Bangalore, India

**Kazim Ahmed Muneer**
Department of Information Technology
Priyadarshini Engineering College
Tirupattur, India

**Bhanu Prakash Lohani**
ASET
Amity University
Greater Noida, India

**Maan Vinod**
Department of Computer Science & Engineering
Mody University of Science & Technology
Lakshmangarh, India

**Marwaha Anu**
Department of Computer Science & Engineering
Sheridan College
Oakville, Canada

**Mehra Ankush**
Department of Computer Science and Engineering
Lovely Professional University
Phagwara, India

**Nijhawan Monika**
New Delhi Institute of Management
Delhi, India

**Pandey Shriyansh**
School of Computer Science Engineering and Technology
Bennett University
Greater Noida, India

**Pradeep Kumar Kushwaha**
ASET
Amity University
Greater Noida, India

**Rakesh Kumar Choudhary**
School of Computer Science Engineering and Technology
Bennett University
Greater Noida, India

**Ramya G.**
Department of Information Technology
National Engineering College
Kovilpatti, India

**Ravindra Kumar Singh Rajput**
G. B. Pant University of Agriculture & Technology
Pantnagar, India

**Soni Rajshree**
Department of Computer Science & Engineering
Mody University of Science & Technology
Sikar, India

**Syed Muzammil Munawar**
Department of Biochemistry
C. Abdul Hakeem College (Autonomous)
Ranipet, India

**Tanwar Sarvesh**
Amity Institute of Information Technology
Amity University
Noida, India

**Tamilselvi S.**
Department of Electronics and Communication Engineering
National Engineering College
Kovilpatti, India

**Vijai C.**
School of Commerce
Vel Tech Rangarajan Dr. Sagunthala R&D Institute of Science and Technology
Chennai, India

Chapter 1

# Basics of blockchain technology and it's underlying principles

*Kafeel Ahmed Muneer, Kazim Ahmed Muneer, and Syed Muzammil Munawar*

## 1.1 INTRODUCTION

Software-Defined Networking (SDN) introduces network abstractions, hence increasing network programmability and flexibility. It enables network administrators and engineers to quickly create and deploy network applications. The SDN controller is the core intelligence centre of the SDN architecture [1], playing an important role in network administration. It provides a comprehensive view of the network, allowing key tasks like link failure monitoring, traffic engineering, and load balancing to be completed more efficiently. Responding rapidly to connection problems is critical for maintaining network performance [2]. These failures are triggered by a variety of factors, including device problems, bottlenecks, cable cutting, and malicious network activity, resulting in significant communication disruptions. Link-failure recovery approaches have been extensively studied across a variety of network paradigms, including classical networks, network function virtualisation, and SDN networks. Effective link failure recovery systems target to detect, recover, and manage network congestion in less than 50 ms [3]. The scholarly literature distinguishes two major techniques to dealing with connection failures: proactive recovery and reactive recovery. Proactive recovery is implementing a recovery strategy before a failure happens, whereas reactive recovery responds to failures as they occur. SDN technology, especially when shared with the OpenFlow protocol, provides a solid framework for developing link failure recovery solutions. Unlike traditional IP networks, SDN's flow tables can be set either reactively or proactively [4]. With a complete image of the network, the SDN controller can intelligently distribute network resources, enhancing network efficiency [5,6], besides, network traffic is routed between switches using flow entries, and it is easily diverted and rescheduled.

### 1.1.1 Architectures based on OpenFlow and SDN

The SDN architecture has three hierarchical planes: application, control, and data. The data plane includes the network's hardware infrastructure, such as routers and switches. The network control plane, located above the

DOI: 10.1201/9781003504559-1

data plane, is at the heart of the SDN architecture, containing a centralised controller in charge of devising and disseminating network policies to all data-plane devices [7]. This controller could be a single server or a cluster of servers. The application plane contains a wide range of programmes, including load balancers, traffic-engineering tools, connection-failure recovery methods, voice over IP services, network monitoring utilities for optimising network operations, and third-party apps. Northbound Application Programming Interfaces (APIs) provide connection between the SDN controller and the application plane, whereas southbound APIs allow communication between data-plane devices and the SDN controller, with the OpenFlow protocol being the most common southbound API [8]. The Open Networking Foundation created OpenFlow, a standard protocol that has been widely adopted by large organisations such as Cisco, Huawei, HP, Facebook, and Google. However, OpenFlow is not the only way to communicate between SDN controllers and southbound APIs. Alternative protocols, such as Forwarding and Control Element Separation (ForCES), are also available and operational. While OpenFlow is widely used because it is compatible with modern Ethernet routers and switches that support flow tables for a variety of networking tasks such as statistical data stream analysis, routing, firewall protection, and subnetting, other protocols, such as ForCES, are also viable options. The primary goal of SDN is to create a more controllable network architecture by enabling logically centralised control and network programmability. ForCES, while less common than OpenFlow, lacks open-source support and may employ disruptive corporate methods. Detailed comparisons of the ForCES and OpenFlow protocols are available for additional investigation. SoftRouter, however less extensively used, is another protocol that facilitates communication between the control and data planes [9,10] (Figure 1.1).

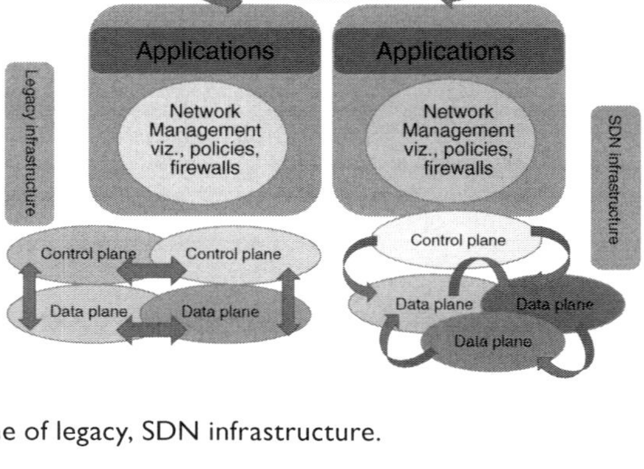


*Figure 1.1* Outline of legacy, SDN infrastructure.

### 1.1.2 SDN link failures

Traditional IP network devices, such as switches, lack programmability due to the locked firmware and closed design enforced by manufacturers. This constraint stifles network innovation because engineers can configure these devices only based on manufacturer standards [11]. SDN transforms network programmability, providing more freedom for creativity and innovation. SDN's programmability enables the development of applications to automate and manage many aspects of network connection, dependability, and flexibility [12]. Network engineers have the capability to employ high-level programming languages for the creation and administration of networks encompassing various devices, topologies, connection-failure management, Quality of Service, and flexible traffic-flow regulations. In this setup, SDN switches serve as forwarding devices, linked to an SDN controller responsible for overseeing both the switches and the network as a whole. Essentially, a switch is dependent on the controller for functionality, suggesting that it cannot operate independently. As a result, the SDN controller continuously analyses switches and makes decisions based on connection or network issues [13]. The SDN controller, with its centralised management of the entire network, enables network virtualisation, resulting in more efficient link-failure recovery procedures. Furthermore, the Path Computation Element protocol and architecture play important roles in network management and path computation [14].

#### *1.1.2.1 Link failure recovery design built on OpenFlow*

The OpenFlow protocol forms the basis for one of the most extensively used link-failure recovery architectures addressed in the literature. This architecture consists of three major modules in the switches and a separate link-failure recovery management module within the controller. In this architecture, the controller takes the initiative in proactively identifying and establishing backup channels within the switches [15]. Meanwhile, the switches are responsible for creating renewal packets to ensure ongoing flow delivery to their destinations while retaining the operation of current backup channels. When the switches detect a failed link, they erase all linked records. When the link is restored, the switches notify the controller, which allows for a recalculation of the optimal path [16].

#### *1.1.2.2 Manager of link failure recovery*

The link malfunction upturn administrator creates substitute channels for the entire switches.

When preparing for a flow request, the controller selects both the primary working and backup paths. Usually, backup channels are established using connection-protection techniques. This computation is carried out on all nodes along a predetermined path to the destination. In the occasion of a connection breakdown, the impacted switch can redirect all traffic to a pre-defined backup path [17].

#### 1.1.2.3 Link detection and removal

Despite switches being pre-configured with alternate flow entries, they are unable to prevent unsuccessful connections from being stored in their flow tables. As a result, all unsuccessful link-flow entries must be removed from the system. To do this, the switch must first determine which port connects it to the failed links before erasing all records associated with that port using the port information. Once these entries are removed, the switches can select alternate paths without relying on the SDN controller's instructions [18].

#### 1.1.2.4 Packet extension

The OpenFlow standard requires each flow entry to include both hard and inactive timeouts, with an idle time of n milliseconds, but here is nix pour interchange detected during this time, the switch's flow table must be cleared [19]. As a result, in the absence of a connection issue, alternate pathways are eliminated after n milliseconds. To address this issue, switches generate renewal packets in a specific format and distribute it periodically, diverting alternate paths away from the currently active path. The switch sending the regenerated packet can uniquely identify it using a defined field, preventing it from reaching its intended destination. If the flow of a failing link is restored, traffic may return to the original link, which is the quickest route. If the switches do not request that the controller recalculate the working path, they will continue to use alternate path entries. To overcome this issue, switches send reinstatement packets for each flow when they reconnect to a previously failed link. The controller then computes and saves the working and alternate paths into the switch's flow tables [20].

#### 1.1.2.5 Taxonomy of SDN relation breakdown revival

The copy on SDN connection breakdown revival solutions categorise them into two main approaches: reactive strategies and proactive processes. Reactive strategies respond to failures as they happen, while proactive methods anticipate failures and implement preventative measures. The section concludes by evaluating the compensation and decompensation of connection breakdown revival. Reactive methods often result in prolonged revival times when a solution is sought following a failure. In contrast, proactive approaches may underutilise resources, since they foresee future failures that may not occur [21].

#### 1.1.2.6 Recovery of proactive link failures

The authors propose a two-pronged approach to mitigating and reducing link failures in wireless mesh networks. Initially, the technique predicts possible connection failures in both the control and data planes of wireless mesh networks. Radio fading, connection noise, node migration, node deafness, and traffic congestion are all potential causes of such failures [22]. To predict these failures, the authors use the signal-to-noise ratio (SNR) as a metric to assess network performance and predict node mobility. In SNR prediction, the duration of the monitoring period plays a crucial role in forecast accuracy. Longer monitoring periods are suggested to enhance prediction accuracy, albeit at the cost of increased delays. The authors employ Kalman filters to smooth out fluctuations in measured SNR and make precise state predictions. The filtered SNR signal is approximately linearly proportional to the distance between the sender and receiver nodes, meaning that changes in SNR values correspond to changes in the distance between the two nodes. Readers can consult supplementary literature for a more detailed treatment of Kalman filters [23,24]. To predict data-plane link failures, the authors compare measured SNR values from the failed node, which was caused by mobility, to patterns of recorded SNR data from the current link. Similar trends indicate an approaching link failure, leading to the detection of a connection breakdown in the data plane, the controller is promptly notified. Leveraging access to numerous sets of SNR data, the controller utilises Support Vector Machines to make accurate predictions. These predictions rely on SNR values collected from neighbouring nodes, enabling the controller to anticipate potential failures with precision. The second component of the proposed approach is to generate an alternate routing path based on node mobility forecasts. This path is chosen based on factors such as shortest path distance, lowest control overheads, and minimal disruption to existing traffic. Before a link fails, the control plane anticipates potential connection losses, calculates alternative pathways, and sends rerouting instructions to affected nodes. The suggested two-layer system is tested using the ns-3 discrete-event simulator, which replicates a network of 50 nodes distributed across 450,000 $m^2$. Furthermore, because computationally difficult processes such as support vector machine and linear programming are performed within the controller, which has enough processing power, the complexity of these operations has no impact on the overall efficiency of SDN or introduces routing delays. Additionally, Stateful Programmable Failure Detection and Recovery (SPIDER) is introduced as a fault tolerant, programmable SDN pipeline for detecting and recovering switch problems quickly. SPIDER, which is based on stateful data planes for speedy failure detection and recovery, works similarly to an OpenFlow pipeline, with periodic link probing to detect switch link failures. Rerouting paths are immediately calculated after detection, regardless of the controller's state or availability [25]. Aside from assisting with rerouting prior to failure, SPIDER includes a method to restore original packet forwarding once the problem has been resolved. The authors explain how to recover from single-link failures via performance and resource sharing. This proactive failure-recovery strategy considers backup path performance and resource utilisation, hence reducing backup flow resource consumption while promoting backup resource utilisation [26]. To enable real-time self-recovery, the system incorporates a revolutionary smart grid recovery mechanism and link failure detecting capabilities. With the release of version 1.1.0, all OpenFlow protocols now support the proactive OpenFlow fast-failover group technique for detecting link failures. This enables switches to dynamically adjust their flow rules based on the status of active ports, enhancing network resilience. The authors' findings indicate that the SDN-based recovery solution is capable of meeting the stringent 4 ms latency requirement for teleprotection applications across both Wi-Fi and LTE networks. This assessment considers elements such as switching time, packet-flow delay, packet loss, and the frequency with which packets fail to meet the teleportation latency requirement due to connectivity issues [27]. Despite the advantages of splitting the control and data planes in SDN, link-failure recovery can be difficult due to intrinsic delays between the breakdown location and the controller. To reduce latency and packet loss, the authors suggest using preset backup paths to instantly reroute disturbed flows to their destination rather than relying on the controller's judgement. They also highlight potential issues with rerouting flows, such as high construction costs for backup channels and the difficulty of changing them once established, necessitating ongoing backup path maintenance to respond to network conditions and alleviate congestion [28]. The authors advocate for a proactive approach to address congestion and connection failures, which involves establishing multiple backup pathways for each link utilising techniques such as source routing, per-hop tagging, and distributing flows across various paths. Their experiments and simulations validate the feasibility of implementing congestion-aware link-failure recovery in SDN, showcasing the potential to mitigate overhead while enhancing network reliability [29]. The authors also present a method for recognising switch failures, highlighting the need for detecting switch failures before they occur in order to minimise potential cascading link-failures induced by switch failures. Treating only link issues after a switch failure may be fruitless, underlining the significance of proactive switch-failure identification [30].

#### *1.1.2.7 Link breakdown revival should be reactivated*

TFLink is a hybrid system designed to address difficulties such as limited Ternary Content Addressable Memory, bandwidth control, and extended recovery durations that have been discovered in previous studies [31]. TFLink operates in three main steps: To begin, it generates backups of the existing network flow rules and links, optimising the overall cost of Ternary

Content Addressable Memory and connection bandwidth to provide for efficient link recovery. These backup links and supplemental flow rules are kept in the controller's global mapping tables, not on the switches [32]. Second, in case of a link failure, TFLink creates new links and flow rules for switches to ensure speedy traffic-control restoration and protection systems based on proven data-traffic restoration and protection procedures for out-of-band networks. To attain carrier-grade quality, TFLink aims for network recovery from disturbances in less than 50 ms. To recover from SDN-based connection failures, it employs a shortest-path quick rerouting approach, which opens up extra paths to relieve network congestion and evenly distribute traffic. It also provides a rapid failover solution for managing controller failures, removing the significant network traffic overhead caused by inter-domain controller synchronisation. Experimental results reveal that TFLink decreases flow modification by around 75% and service interruption duration by roughly 50% when compared to standard SDN baseline solutions following a single link loss. Furthermore, the authors present BOND, a versatile link failure and recovery solution that produces backup paths and assigns backup rules to forwarding switches based on current flows. BOND uses a global hash table to precisely identify backup channels, minimising potential congestion and improving failure recovery. Extensive real-world testing has confirmed the efficiency and practicality of BOND, showcasing its capability to calculate all backup paths and deploy backup rules seamlessly in the event of a single breakdown revival [33].

## 1.2 SUMMARY AND FUTURE ENHANCEMENTS

The offered technique incorporates both proactive and reactive elements while taking into account the advantages and disadvantages of the available options. It recognises that most failures stem from either a faulty link between switches or a malfunctioning switch, so it prioritises preventive measures. However, some strategies presented in the literature combine features of reactive and proactive techniques [34].

## 1.3 CONCLUSIONS

The chapter illustrated how SDN and OpenFlow technology frameworks allow for the rapid discovery and recovery of connection failures. It compared OpenFlow to many alternative protocols for managing communication between control and data planes. It also provided the criteria for evaluating potential solutions. According to the study's findings, SDN controllers have the ability to make intelligent traffic-flow decisions in the case of a link loss due to their comprehensive view of the network.

Furthermore, SDN technology enables the quick construction of modules like the link-failure-recovery management module, which computes backup flow configurations. A typical OpenFlow-based link-failure-recovery system consists of four components: the control plane's failure-recovery manager, a link module for detection and removal, packet renewal, and path module reinstatement. The latter three modules rely on data-plane switches. This perfect architecture has allowed academics to develop novel strategies for recovering from connection failures in SDN [35].

## 1.4 FUTURE RESEARCH PROJECTS

Finally, SDN connects failure-recovery themes and approaches, revealing a number of challenging areas that demand further investigation and resolution as future research directions. Link synchronisation failed [36]. Addressing instances in which a link is down in one direction but up in the other, as is prevalent in fibre optic networks. Mechanisms for synchronised session-state management between endpoints should be built to promptly detect and share failure information, as well as data plane programmability [37]. While SDN is primarily concerned with controller plane programmability, improving data-plane programmability has the potential to alter network management. Further research is essential to fully understand the implications of enhanced programmability in the data plane, such as potential improvements in packet security, compression, and decision-making latency reduction. Intelligent load balancing is necessary for good link recovery to prevent congestion when adding new channels. To maintain network performance, any link-failure-recovery plan should prioritise congestion avoidance, as indicated in Gradual SDN implementation [38]. Transitioning to SDN gradually can be challenging, especially in hybrid networks. Additional study is needed to determine which network components should be updated to SDN capability first, as well as to discover connection difficulties in hybrid SDN systems, particularly as more organisations choose for staged SDN deployment with end-user-friendly APIs [39]. While the SDN control plane is primarily intended for network managers, providing user-friendly APIs enables end users to access on-demand services such as traffic filtering or bandwidth allocation to improve performance. Despite concerns about security and trust, developing user-friendly APIs between client applications and the control plane is deemed vital. To summarise, overcoming these issues and exploring these research topics will be crucial for improving SDN link-failure-recovery and reaching the full potential of SDN technology in network management and administration [40].

**Conflict of Interest:** The authors declare that there are no conflicts of interest.

## ACKNOWLEDGEMENT

The authors are thankful to Dr. Sarvesh Tanwar, Professor at Amity University, Noida, India, for their encouragement, providing the necessary facilities, and support in carrying out the work.

## REFERENCES

1. Ramon Casellas, Raül Muñoz, Ricardo Martínez, Ricard Vilalta, Lei Liu, Takehiro Tsuritani, Itsuro Morita, Víctor López, Oscar González de Dios, Juan Pedro Fernández-Palacios, SDN orchestration of OpenFlow and GMPLS flexi-grid networks with a stateful hierarchical PCE, *J. Opt. Commun. Netw.* 7 (1) (2015) A106–A117.
2. Raül Muñoz, Ricard Vilalta, Ramon Casellas, Ricardo Martínez, Frederic Francois, Mayur Channegowda, Ali Hammad, Shuping Peng, Reza Nejabati, Dimitra Simeonidou, et al., Transport network orchestration for end-to-end multilayer provisioning across heterogeneous SDN/OpenFlow and GMPLS/PCE control domains, *J. Lightwave Technol.* 33 (8) (2015) 1540–1548.
3. Sgambelluri Andrea, Fillippo Paolucci, Alessio Giorgetti, Fillippo Cugini, Piero Castoldi, SDN and PCE implementations for segment routing, in: *2015 20th European Conference on Networks and Optical Communications-(NOC)*, IEEE, 2015, pp. 1–4.
4. Francesco Paolucci, Filippo Cugini, Alessio Giorgetti, Nicola Sambo, Piero Castoldi, A survey on the path computation element (PCE) architecture, *IEEE Commun. Surv. Tutor.* 15 (4) (2013) 1819–1841.
5. Sanjeev Singh, Rakesh Kumar Jha, A survey on software defined networking: Architecture for next generation network, *J. Netw. Syst. Manage.* 25 (2) (2017) 321–374.
6. Qadeer Yasin, Zeshan Iqbal, Muhammad Attique Khan, Seifedine Kadry, Yunyoung Nam, Reliable multipath flow for link failure recovery in 5G networks using SDN paradigm, *Inf. Technol. Control* 51 (1) (2022) 5–17.
7. Malcolm Betts, S. Fratini, N. Davis, D. Hoods, R. Dolin, M. Joshi, Z. Dacheng, SDN architecture, issue 1, open networking foundation, Technical report, ONF TR-502, 2014.
8. Thabo Semong, Kun Xie, Efficient load balancing and multicasting for uncertain-source SDN: Real-time link-cost monitoring, in: *Computer Science on-Line Conference*, Springer, 2018, pp. 178–187.
9. Arsalan Tavakoli, Martin Casado, Teemu Koponen, Scott Shenker, Applying NOX to the datacenter, in: HotNets, 2009.
10. Ulas C. Kozat, Guanfeng Liang, Koray Kokten, Janos Tapolcai, On optimal topology verification and failure localization for software defined networks, *IEEE/ACM Trans. Netw.* 24 (5) (2015) 2899–2912.
11. Lingxia Liao, Victor C.M. Leung, LLDP based link latency monitoring in software defined networks, in: *2016 12th International Conference on Network and Service Management (CNSM)*, IEEE, 2016, pp. 330–335.

12. Farzaneh Pakzad, Marius Portmann, Wee Lum Tan, Jadwiga Indulska, Efficient topology discovery in software defined networks, in: *2014 8th International Conference on Signal Processing and Communication Systems (ICSPCS)*, IEEE, 2014, pp. 1–8.
13. Ke Bao, John D. Matyjas, Fei Hu, Sunil Kumar, Intelligent software-defined mesh networks with link-failure adaptive traffic balancing, *IEEE Trans. Cogn. Commun. Netw.* 4 (2) (2018) 266–276.
14. Sigurd I. Aanonsen, Geir Nævdal, Dean S. Oliver, Albert C. Reynolds, Brice Vallès, et al., The ensemble Kalman filter in reservoir engineering: a review, *Spe J.* 14 (03) (2009) 393–412.
15. Steve R. Gunn, Support vector machines for classification and regression, *ISIS Tech. Rep.*, 14 (1) (1998) 5–16.
16. Carmelo Cascone, Luca Pollini, Davide Sanvito, Antonio Capone, Brunilde Sanso, SPIDER: Fault resilient SDN pipeline with recovery delay guarantees, in: *2016 IEEE NetSoft Conference and Workshops (NetSoft)*, IEEE, 2016, pp. 296–302.
17. Ying Wang, Sixiang Feng, Hantao Guo, Xuesong Qiu, Hengbin An, A single-link failure recovery approach based on resource sharing and performance prediction in SDN, *IEEE Access* 7 (2019) 174750–174763.
18. Abdullah Aydeger, Nico Saputro, Kemal Akkaya, Selcuk Uluagac, SDN-enabled recovery for smart grid teleprotection applications in post-disaster scenarios, *J. Netw. Comput. Appl.* 138 (2019) 39–50.
19. Liaoruo Huang, Qingguo Shen, Wenjuan Shao, Congestion aware fast link failure recovery of SDN network based on source routing, *TIIS* 11 (11) (2017) 5200–5222.
20. Zhaogang Shu, Jiafu Wan, Jiaxiang Lin, Shiyong Wang, Di Li, Seungmin Rho, Changcai Yang, Traffic engineering in software-defined networking: Measurement and management, *IEEE Access* 4 (2016) 3246–3256.
21. Rahim Masoudi, Ali Ghaffari, Software defined networks: a survey, *J. Netw. Comput. Appl.* 67 (2016) 1–25.
22. Ian F. Akyildiz, Ahyoung Lee, Pu Wang, Min Luo, Wu Chou, A roadmap for traffic engineering in SDN-OpenFlow networks, *Comput. Netw.* 71 (2014) 1–30.
23. Binghao Yan, Qinrang Liu, JianLiang Shen, Dong Liang, Bo Zhao, Ling Ouyang, A survey of low-latency transmission strategies in software defined networking, *Comp. Sci. Rev.* 40 (2021) 100386.
24. Muhammad Yunis Daha, Mohd Soperi Mohd Zahid, Babangida Isyaku, Abdussalam Ahmed Alashhab, CDRA: a community detection based routing algorithm for link failure recovery in software defined networks, *Int. J. Adv. Comput. Sci. Appl.* 12 (11) (2021) 712–722.
25. Ali Malik, Benjamin Aziz, Mo Adda, Chih-Heng Ke, Optimisation methods for fast restoration of software-defined networks, *IEEE Access* 5 (2017) 16111–16123.
26. Tao Hu, Peng Yi, Julong Lan, Yuxiang Hu, Penghao Sun, FTLink: efficient and flexible link fault tolerance scheme for data plane in Software-Defined Networking, *Future Gener. Comput. Syst.* 111 (2020) 381–400.

27. Paulo Cesar da Rocha Fonseca, Edjard Souza Mota, A survey on fault management in software-defined networks, *IEEE Commun. Surv. Tutor.* 19 (4) (2017) 2284–2321.
28. Vuthu Muthumanikandan, Chinnaiah Valliyammai, A survey on link failures in software defined networks, in: 2015 Seventh International Conference on Advanced Computing (ICoAC), IEEE, 2015, pp. 1–5.
29. Stewart Bryant, Stefano Previdi, Mike Shand, A framework for IP and MPLS fast reroute using not-via addresses, RFC 6981, 2013.
30. Srihari Nelakuditi, Sanghwan Lee, Yinzhe Yu, Zhi-Li Zhang, Chen-Nee Chuah, Fast local rerouting for handling transient link failures, *IEEE/ACM Trans. Netw.* 15 (2) (2007) 359–372.
31. Alia Atlas, Alex Zinin, Basic specification for IP fast reroute: loop-free alternates, Technical report, RFC 5286, September 2008.
32. Jue Chen, Jinbang Chen, Fei Xu, Min Yin, Wei Zhang, When software defined networks meet fault tolerance: A survey, in: *International Conference on Algorithms and Architectures for Parallel Processing*, Springer, 2015, pp. 351–368.
33. Yong Li, Min Chen, Software-defined network function virtualization: a survey, *IEEE Access* 3 (2015) 2542–2553.
34. Rashid Mijumbi, Joan Serrat, Juan-Luis Gorricho, Steven Latré, Marinos Charalambides, Diego Lopez, Management and orchestration challenges in network functions virtualization, *IEEE Commun. Mag.* 54 (1) (2016) 98–105.
35. Shiming He, Kun Xie, Xuhui Zhou, Thabo Semong, Jin Wang, Multi-source reliable multicast routing with QoS constraints of NFV in edge computing, *Electronics* 8 (10) (2019) 1106.
36. Andrea Sgambelluri, Alessio Giorgetti, Filippo Cugini, Francesco Paolucci, Piero Castoldi, OpenFlow-based segment protection in ethernet networks, *J. Opt. Commun. Netw.* 5 (9) (2013) 1066–1075.
37. Nick McKeown, Tom Anderson, Hari Balakrishnan, Guru Parulkar, Larry Peterson, Jennifer Rexford, Scott Shenker, Jonathan Turner, OpenFlow: enabling innovation in campus networks, *ACM SIGCOMM Comput. Commun. Rev.* 38 (2) (2008) 69–74.
38. Ali Akbar Neghabi, Nima Jafari Navimipour, Mehdi Hosseinzadeh, Ali Rezaee, Load balancing mechanisms in the software defined networks: a systematic and comprehensive review of the literature, *IEEE Access* 6 (2018) 14159–14178.
39. Andrew Goodney, Saurabh Kumar, Akshay Ravi, Young H. Cho, Efficient PMU networking with software defined networks, in: *2013 IEEE International Conference on Smart Grid Communications (SmartGridComm)*, IEEE, 2013, pp. 378–383.
40. Likun Wang, Lin Yao, Zichuan Xu, Guowei Wu, Mohammad S. Obaidat, CFR: A cooperative link failure recovery scheme in software-defined networks, *Int. J. Commun. Syst.* 31 (10) (2018) e3560.

Chapter 2

# Revolutionizing network architecture

## The promise and potential of software-defined networking

*Vinod Maan and Rajshree Soni*

## 2.1 SOFTWARE-DEFINED NETWORK

The emerging digital world, propelled by the Internet, is revolutionizing society. Almost everything has been linked to the Internet via the Internet of Things. The network's setup and administration, where the bearer and control planes are closely related, is computationally very complex and expensive with the traditional IP network system.

Numerous shifts have been observed in network traffic patterns because of the increasing prevalence of cloud-based applications. It has always been difficult to accommodate intricate network patterns and adjust to frequent changes in a typical network [1]. The idea of SDN aims to make network management tasks easier.

The majority of businesses are currently transitioning to SDN infrastructures that are a significant change from the architecture of conventional networks. SDN boosts the flexibility and efficiency of networks by centralizing control. SDNs control network traffic and communicate with the underlying hardware infrastructure through the use of controllers that are software-based or application programming interfaces (APIs) [2].

In SDN, two planes are involved.

1. Control Plane
2. Bearer plane

**Control Plane**: The network segment in charge of managing data packet forwarding, or the transfer of data from one location to another, is called the control plane. For instance, any of the functions of the control plane are thought to be facilitating the development of a routing table. Network paths are identified by routers using a wide range of available protocols, and are then subsequently preserved in routing tables. The control plane uses a wide range of protocols, including:

 DOI: 10.1201/9781003504559-2

- Routing protocols (OSPF, IS-IS, and BGP)
- Procedures for network management (SNMP)
- Protocols at the application layer (HTTP and FTP)

SDN is used to a great extent by these protocols to establish virtual networks and control their traffic [3]. SDN-enabled virtual networks play a key role in enterprise-level data traffic management. They make it possible for organizations to:

- Traffic segmentation
- Set critical data flows in order of importance.
- Separate traffic coming from various network segments.

**Bearer Plane**: The bearer plane, as presented in Figure 2.1, is the component that actively forwards the packets, as opposed to the control plane, which determines how they are supposed to be forwarded. The bearer plane is also referred to as the forwarding plane. Here are a few instances of bearer planes:

- Ethernet configuration
- Wireless networks
- Cellular systems
- Satellite-based connectivity

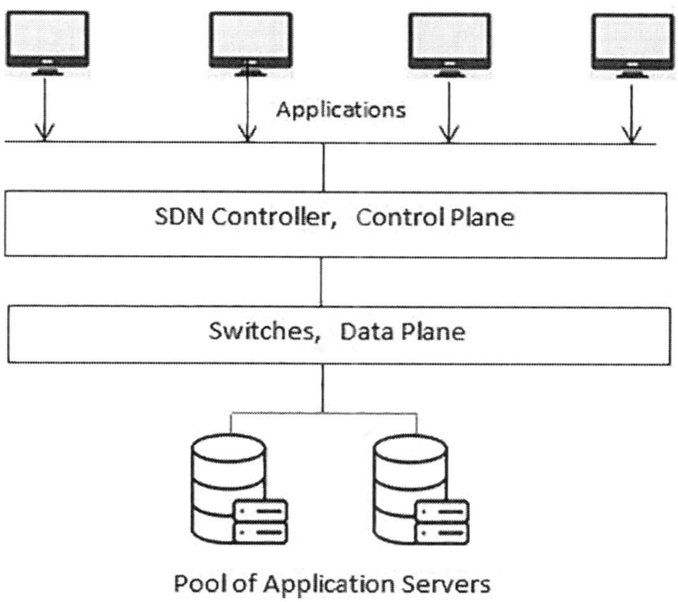


*Figure 2.1* Software defined networking (SDN).

Virtualized networks, such as those built with software-defined networks or virtual private networks, can also be included in bearer planes. Bearer planes may additionally include specialized networks, such as industrial control systems or the Internet of Things [2].

Organizations can securely and swiftly move data between systems with the help of bearer planes. A bearer plane, for instance, can facilitate data transfer between a local system and an application running in the cloud. Organizations that need to swiftly move massive amounts of data or that need to access data from several systems may find this functionality useful.

SDN is important for many reasons:

1. **Flexibility and Liveliness**: SDN isolates control and bearer planes, so that traffic flows of the network can be centrally controlled. This separation allows networks to be more flexible and able to adapt to changing requirements, by dynamically setting up and reconfiguring the network's resources.
2. **Automation**: SDN uses programmable interfaces to facilitate the automation of network management tasks. This automation facilitates faster provisioning, minimizes human configuration errors, and improves resource efficiency.
3. **Scalability**: Conventional networking topologies can be intricate and challenging to grow. By offering a central view and control of network resources, SDN streamlines network management and makes it simpler to scale networks up or down as required.
4. **Cost Savings**: SDN can cut down costs by separating network intelligence from physical devices, which can save money on both capital and operating expenses (OPEX and CAPEX). It eliminates the need for proprietary networking equipment, permits the use of less expensive commodity hardware, and automates the reduction of operating expenses.
5. **Better Network Visibility and Analytics**: By centralizing monitoring and analytics, SDN offers improved insight into network performance and traffic. Because of this visibility, network administrators can find and fix problems faster, which enhances the performance and dependability of the network.
6. **Support for New Services and Applications**: Because of its programmable interfaces, SDN facilitates the quick deployment of additional network facilities and applications. Organizations can innovate and launch new services more quickly because of this flexibility, which gives them a market advantage.
7. **Security**: By enabling more precise control over network traffic flows, SDN can improve network security. At the network level, policies can be dynamically enforced to lower the attack surface and enhance the overall security posture.

8. **Multi-Tenancy Support**: SDN makes it conceivable for cloud operators and service providers to create virtualized network slices that can accommodate several tenants or customers on a single physical infrastructure. This multi-tenancy support facilitates more effective service delivery and maximizes the use of resources.

Ultimately, SDN is an essential technology for modernizing and optimizing network infrastructure because it provides a more flexible, scalable, and economical approach to network administration [1].

There are numerous significant ways that SDN differs from conventional networking.

1. **Control Plane Separation**: In conventional networking, network apparatus like switches and routers have a close integration between control planes, which take decisions regarding traffic forwarding, and the bearer plane, which actually forwards traffic. These planes are decoupled in SDN. The control plane is then relocated to a central controller, which decides how traffic should be routed and offers a worldwide perspective of the network. The bearer plane, which is constructed of forwarding devices, merely executes the controller's instructions [3].
2. **Programmability**: Using open Application Programming Interfaces, SDN makes networks programmable, enabling administrators to automate network orchestration, management, and arrangement. Changes to configuration in traditional networking are usually made manually on each device, which can be laborious and prone to mistakes.
3. **Centralized Management**: Network supervisors can define policies and configurations for the whole network from only one point of control due to SDN's centralized management and control of the network. Because management tasks are dispersed among several devices in traditional networking, it is more difficult to preserve uniformity and visibility throughout the network.
4. **Dynamic Adaptation**: SDN provides dynamic adaptation to shifting traffic patterns and network conditions. SDN controllers can dynamically modify network configurations and routing choices in response to events and application requirements that arise in real time by centralizing control and utilizing programmability. Changes in traditional networks are often implemented manually, which makes it more difficult for them to quickly adapt to changing needs.
5. **Hardware Independence**: SDN frees enterprises from vendor lock-in by separating network intelligence from the hardware that powers it. This enables them to use commodity hardware. Conventional networking is frequently dependent on proprietary hardware and protocols, which restricts flexibility and raises expenses.

6. **Service Orchestration**: By facilitating the quick deployment and scalability of network resources and services, SDN makes service orchestration easier. In contrast to traditional networking, which can involve complicated and time-consuming service deployment, SDN makes it easier to deliver new services and applications through programmable interfaces and centralized control.
7. **Granular Policy Enforcement**: Through programmable control, SDN provides traffic management and granular policy enforcement. Administrators have the ability to create policies and apply them uniformly throughout the network based on particular standards like application type, user identity, or security requirements. Conventional networking might not offer the flexibility and granularity required for efficient policy enforcement.

Compared to traditional networking architectures, SDN provides more flexibility, agility, automation, centralized management, and programmability; as such, it is a good fit for contemporary network environments that are defined by dynamic workloads, changing applications, and rising scalability and efficiency requirements.

## 2.2 THE SOFTWARE-DEFINED NETWORK'S CONSTITUENTS

As mentioned earlier, SDN architecture divides the control and data components of traditional networks. This makes it possible for companies to design computer and networking systems that make the most of network hardware by utilizing software-based technologies. Figure 2.2 lists the principal components of the SDN architectural design.

1. **Application Component**:

   Applications interacting with the controller through Application Programming Interfaces constitute the application component. Additionally, this part creates a summary of the network status, as illustrated in Figure 2.3, while notifying the controller regarding the intended network behavior and resources required. The application layer additionally obtains information from the controller layer and applies it to its decision-making in order to accomplish application goals.

   Applications include business procedures for data center operations, network management, and analytics. An analytics program, for example, can be set up to identify unusual activity and strengthen network security.

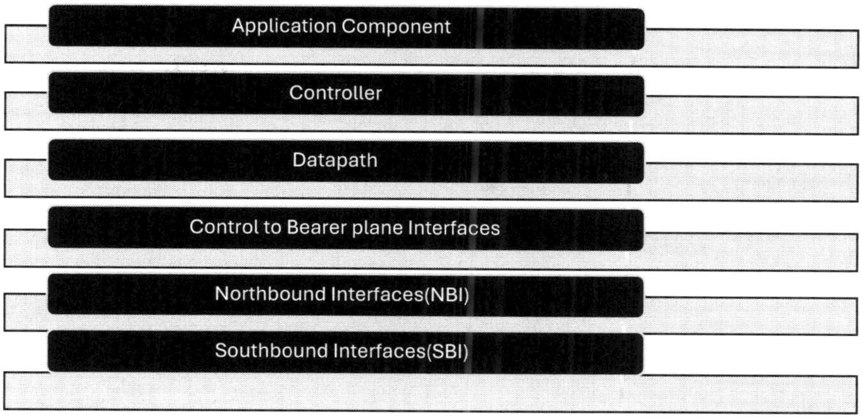


*Figure 2.2* Essential software defined network architectural elements.

2. **Controllers**: The controller component uses logic to process the prerequisites and instructions it receives from the application component before forwarding them to the networking layer. This essential part of the SDN architecture enables network policy enforcement, centralized supervision and management, and automation in both physical and virtual environments. The controller is also responsible for obtaining hardware-layer information about the state and well-being of the network and communicating it to the application component. This enables the application component to generate an abstract network overview that is rich in events and statistics.
3. **Datapaths**: Through the datapath component, users are able to keep an eye on and manage the way that the hardware layer forwards and processes information. This layer is composed of a traffic-forwarding module, a control-data-plane interface agent, and possibly some network traffic-processing modules.

   SDN data paths may be present in one or more instances on a single network device. In a similar vein, multiple devices can have the same SDN datapath defined. This part can also help with virtualization or datapath slicing, logical-to-physical mapping, shared hardware management, and non-SDN networking compatibility.
4. **Control to Data-Plane Interface**: With the help of the control-data-plane interface, the controller and datapath components can communicate with one another. Reporting network statistics, alerting users to significant occurrences, and allowing forwarding operation programming are some of its features. The best SDN solutions contain control-data-plane interface components that are open, interoperable, and vendor-neutral.

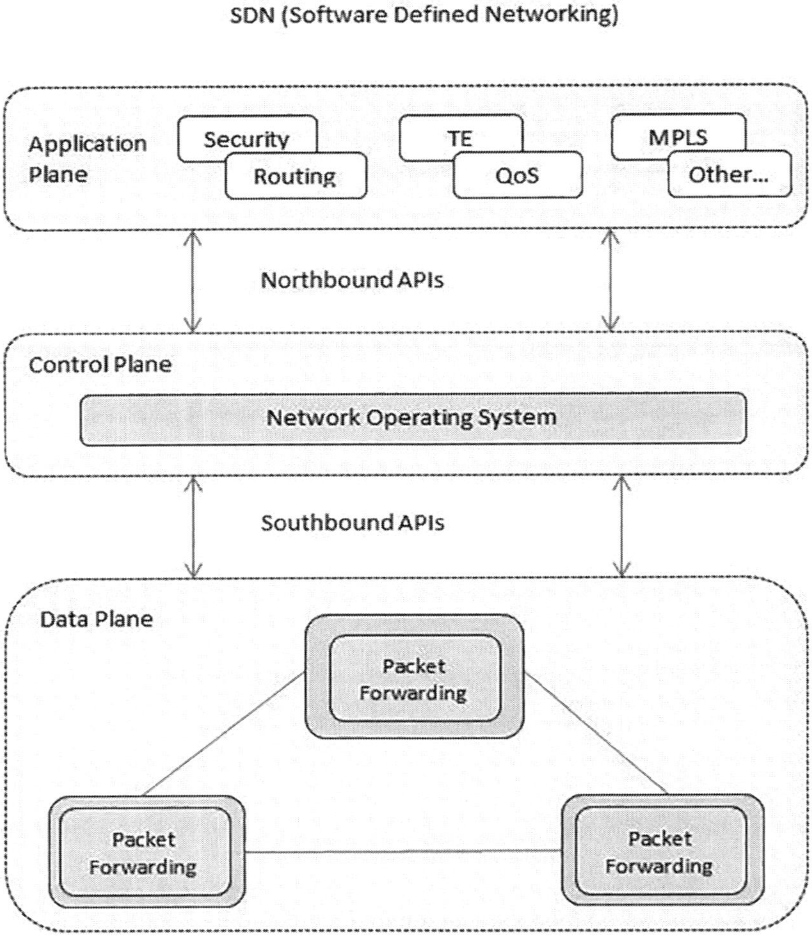


*Figure 2.3* Components of SDN.

5. **Northbound Interface (NBI)**: Data is relayed between the application, controller, and policy layers via the NBI. Regardless of latitude (abstraction) and longitude (functionality), this component usually offers an abstract representation of the network and enables the explicit expression of behavior and requirements.
6. **Southbound Interface (SBI)**: The SBI relays data between the controller component and different hardware devices connected to the network, such as switches, routers, access points, and hardware firewalls. In this section, network concepts are further subdivided into

more precise technical specifications suitable for the lower tier of the architecture. SBIs facilitate communication between higher-level and lower-level network components, such as physical and virtual switches, routers, and nodes. In order to view the network topology, ascertain network flow, and respond to NBI requests. For example, routers rely on the SBI.

## 2.3 SDN APPLICATIONS

There are numerous uses for SDN in different networked environments. Through the separation of the control plane from the bearer plane, programmable networks facilitate personalized control and the chance to do away with intermediaries and make the creation and implementation of new network protocols and services easier. The situations for which SDN clarifications have been suggested or put into practice are depicted in Figure 2.4.

1. **Enterprise Networking:** Huge networks are recurrently managed by enterprises, which also have stringent security as well as performance standards. Moreover, the requirements, features, and user base of various enterprise environments can vary greatly. One particular instance of an enterprise network would be a university network. Because resource allocation and security are more challenging in an academic setting where numerous connecting devices are ephemeral

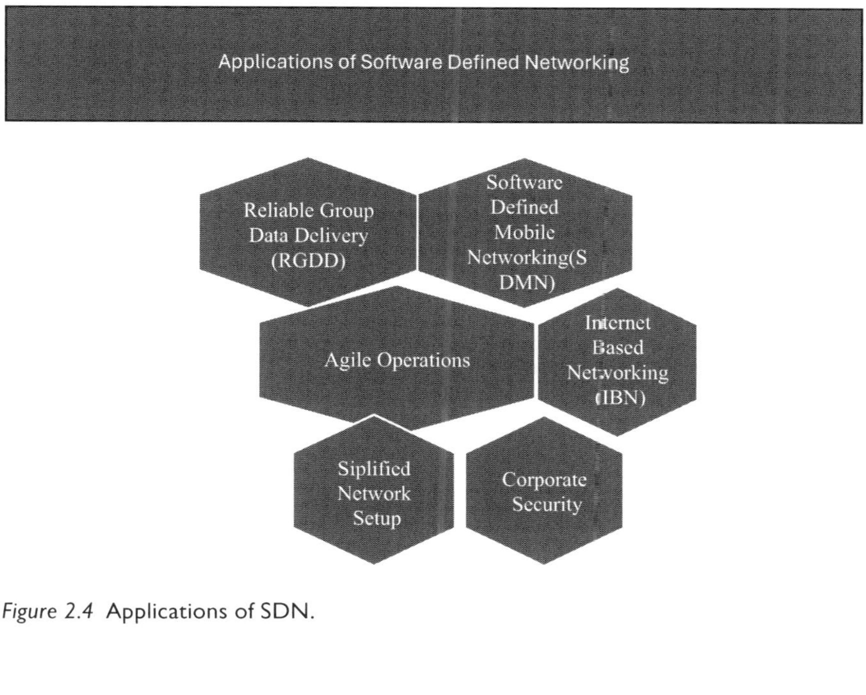


*Figure 2.4* Applications of SDN.

and not under the university's control. Universities also frequently have to support experimental protocols and research testbeds.

In enterprise settings, proper management is essential, and SDN can be considered to programmatically implement and modify network rules, in addition to aiding with monitoring network traffic and adjusting network functionality.

Furthermore, by eliminating middleboxes and incorporating their functionality into the network controller, SDN can be used to streamline the network. SDN has been used to implement middlebox functionality, some notable examples of which are firewalls, load balancers, and network access control [4–6]. SDN can be used to offer unified control and management for more complicated middleboxes that have features (like deep packet inspection) that cannot be employed directly without causing performance degradation [7].

2. **5G**: For cutting-edge uses like fleet management, smart farming, autonomous vehicles, smart cities, and the Industrial Internet of Things, wireless Internet is quickly gaining popularity. This necessitates a standard that enables throughput at speeds never before possible. The quickly developing telecommunications sector is heading toward 5G, the newest standard for cellular connectivity, to meet this need.

   Globally, 5G technology has been gradually introduced since 2019. But this technology's full potential hasn't yet been investigated. It is anticipated that building 5G infrastructure based on SDN will result in fast, open-source networks that will revolutionize the global telecom sector. Businesses should be able to resolve connectivity problems, strengthen network security, diminish latency at a reasonable cost, and improve users' inclusive experience with the aid of SDN-enhanced 5G networks.
3. **Software-Defined Mobile Network (SDMN)**: This kind of arrangement integrates network function virtualization, cloud computing, and SDN in mobile networking environments. It is expected that SDMN will lead the shift from incompatible and static legacy mobile networks to scalable and dynamic connectivity planes. This will be achieved by splitting data and control planes, which are fundamental to SDN. It is essentially expected that SDMN will manifest as an addition of SDN architecture, augmented with functionalities most unique to mobile networks [8].
4. **Intenet-Based Networking (IBN)**: The complexity of connectivity architecture rises due to the exponential growth of new kinds of network-enabled devices, distributed applications, and users that go along with it. Intent-based networking (IBN) can help with that. It is anticipated that IBN will convert manual, hardware-focused networks into controller-led settings that effectively capture user intent and translate it into policies that can be automatically applied consistently

throughout networks. This would allow the network to continuously monitor and adjust performance in order to guide operations toward the intended results.

IBN will take the shape of a closed-loop system that intercepts business intent and converts it into network-actionable policies. Pan-network automation then assists in installing these policies throughout physical and virtual network arrangements. Ultimately, the network is constantly monitored by machine learning and analytics to confirm that the right intent is being applied and to achieve the targeted business results. SDN will be used by IBN to create a centralized network activity control component [9]. IT teams will be able to view the network as a whole, as a cohesive unit as a result. Controller-led networks would make holistic digital transformation much less difficult in areas like access, data centers, WAN, and cloud.

5. **Software-Defined Wide Area Network (SD-WAN)**: SD-WAN is a type of virtual network construction. Companies use LTE, MPLS, broadband Internet, and other transport services in any combination to safely link applications and customers through SD-WAN. SD-WAN improves application performance and user experience by securely directing traffic across the network using a centralized control component. SD-WAN is controlled by using SDN concepts [10]. By separating the control and management components, users can use commercially available leased lines to partially or fully replace MPLS lines, reducing expenses. Some estimates place the cost-effectiveness of SD-WAN at up to 2.5 times that of traditional wide area networks. Moreover, configuration and monitoring are made simpler since network administration is no longer reliant on hardware.

   Even though SD-WAN has been around for a while, it's anticipated that related technologies will be bundled in new ways. New approaches to network aggregation and management, along with dynamic bandwidth sharing among endpoints, are anticipated outcomes of this. SD-WAN is anticipated to gain traction in the near future as more companies digitize their operations to boost productivity and agility [11].
6. **Software-Defined Local Area Network (SD-LAN)**: In data center settings, software-defined local area networks (SD-LANs) make use of these principles. The objective of this network architecture is to improve the cost-effectiveness, scalability, flexibility, and adaptability of wired and wireless access networks. Companies from all industry verticals are operating their functions over the Internet. Maintaining the network's functionality is crucial to them. By separating the hardware- and software-based network layers, SD-LAN rises to the occasion and creates an architecture that is governed by policy and application [12]. The ability of SD-LAN to create self-organizing,

scalable, and easily integrated local area networks under central management is anticipated to drive its demand. Future plans for SD-LAN include the deployment of efficient cloud management systems and wireless connectivity that does not require a physical controller.

7. **Reliable Group Data Delivery (RGDD)**: Effectively managing latency and maintaining uptime are essential for operating a prosperous virtual business in the post-COVID-19 era. Using distributed apps that run across data centers is one method to strengthen redundancy measures and combat latency. For the purposes of load balancing, fault resilience, and bringing information closer to users, these applications replicate and synchronize data. For this to happen, data must be reliably delivered from one server to several clients. Reliable group data delivery is the name given to this data delivery model. It is anticipated that more businesses will use SDN switches for reliable group data delivery. This is to ensure rules that permit information to be forwarded to multiple ports at once can be implemented via SDN. To ensure robust reliable group data delivery, for example, a centralized controller can be configured to create forwarding trees that take load status and network congestion into account [13].
8. **Agile Operations**: In the fast-paced business world, all it takes is one mistake to lose a valuable client to a rival. A state-of-the-art operational methodology is therefore imperative. SDN allays this worry by streamlining network management and enhancing the agility of business networks [14]. The division of forwarding functions results in direct programmability, which enables this. Additionally, SDN increases organizational agility by expanding a company's use of dynamic load balancing. This makes it easier to adjust traffic flow to changes in demand and consumption.
9. **Simplified Network Setup**: It is anticipated that remote work will remain popular well into the upcoming years. This necessitates streamlining the processes involved in network configuration for a variety of scenarios, such as the onboarding of new employees, setting up a new office, and increasing client connectivity. By enabling users to create automated programs, SDN can configure, optimize, and secure network resources as needed, expedites these procedures [15]. Additionally, this architecture makes network design and operation simpler by substituting open controllers for vendor-specific protocols. Micro-segmentation is another tool that SDN makes possible for reducing network complexity. Finally, with the aid of SDN, consistency is created across network architectures, including private, public, hybrid, and multicloud environments.
10. **Cutting-Edge Corporate Security**: Cybersecurity is predicted to be crucial in upcoming years as businesses reevaluate their attack surfaces in response to new-fangled susceptibilities brought on by remote

work. SDN gives businesses more compact and accurate control over network security, which helps to strengthen their security posture. For example, SDN enables network supervisors to centrally implement strategies for various workload types and network segments. Other applications of SDN in cybersecurity include worm propagation prevention, botnet attack prevention, and DDoS attack detection and mitigation. Moving target defense algorithms, which sporadically conceal or alter the essential characteristics of networks to reduce their attack surface, can also be implemented with the assistance of SDN controllers [16].

11. **Network Monitoring and Intelligence**: The most useful layer of data center is the network layer, which can be abstracted with the use of modern SDN technologies. Network architectures now handle more data than ever before and are incredibly complex. This implies that understanding what is happening in the surroundings is essential. Are there problems with packet transmission on a port? If we have a heterogeneous network architecture, what will happen? Alternatively, is the network architecture heavily virtualized and are we moving a lot of data? All of these challenges are lessened if our network cognitive ability and monitoring layer is robust. Nevertheless, if we incorporate these technologies into the SDN architecture, we also reap the rewards and get genuine understanding. Network monitoring and intelligence technologies can also include port configurations, traffic flow, alerting, optimization, and hypervisor integration. We will be able to keep track of network traffic between the data center and the cloud ecosystem with the aid of these agile system types [17].
12. **Security Services**: There are particular virtual services operating at the network layer that are supported by the current virtualization ecosystem. It refers to integrating NFV-like functionality into SDN platforms. With the help of this kind of network security, you can reduce risks and respond to incidents quickly in a truly proactive environment. Every second counts when there is a violation because it can stop the attack. Determining the attack's source and making sure that no other network components are under attack is also crucial. We will witness an increase in the attacks and more complex, advanced persistent threats as organizations in the modern era become even more digitalized and the network layer becomes even more important. You can create an environment that is more proactive, responsive, and change-adaptive by integrating predominant offerings within the SDN layer [18].

## 2.4 CASE STUDIES

The following case studies draw attention to the application of SDN along with its positive benefits in an assortment of industries:

1. **Implementation of SDN by Google**: Google's "Jupiter" adoption of SDN has positioned it as a leader in the field. SDN concepts are used by Google's Jupiter network to connect its data centers all over the world. Performance, flexibility, and cost-efficiency are all increased by Google's ability to centrally manage and optimize traffic throughout its extensive linkage by severing the control plane from the bearer plane.
2. **The SDN Transformation of AT&T**: AT&T started a huge SDN alteration project to revamp its network assistance and the infrastructure. AT&T sought to automate network provisioning, increase scalability, and improve agility in order to satisfy its customers' increasing demands through the implementation of SDN technologies. AT&T has been able to lower operating costs and offer new services more quicker and with greater efficiency because of this transformation.
3. **Tokyo University's SDN Campus Network**: In order to handle the growing demand for bandwidth and meet the various needs of its users, the University of Tokyo put in place an SDN-based campus network. The university was able to gain concentrated oversight of network practices and dynamic resource allocation based on customer demands by implementing SDN. Better resource usage, enhanced network performance, and easier network administration were the outcomes of this.
4. **The SDN for Scientific Research at the Netherlands eScience Center**: An SDN solution was put into place by the Netherlands eScience Center to assist scientific research endeavors that necessitate extensive data processing and cooperation. To support their unique research workflows, researchers can use SDN to optimize data transfers, build custom network topologies, and dynamically provision network resources. This has expedited scientific discoveries and made it easier for researchers to collaborate.
5. **SD-WAN from Verizon for Business Connectivity**: In an attempt to strengthen enterprise connectivity and give companies greater autonomy and oversight over their network infrastructure, Verizon enacted SD-WAN solutions. By means of intelligent traffic routing over the most efficient paths, businesses can cut costs, improve network security, and optimize application performance by leveraging SD-WAN technology. Verizon's SD-WAN solution has given companies the flexibility to adapt to digital transformation and adjust to shifting network necessities.

These case studies highlight the various uses and advantages of SDN for a range of industries, including telecommunications, enterprises, educational institutions, research groups, and tech giants.

## 2.5 CHALLENGES OF SOFTWARE-DEFINED NETWORK

Although the SDN initiative has great potential to change and enhance existing networks, it is still in its early phases of tackling the many challenges spanning multiple disciplines. Specifically, a number of urgent issues pertaining to the distributed control of SDNs demand our particular attention. These concerns include the following: (1) Interoperability; (2) Scalability; (3) Reliability; (4) Consistency; (5) Monitoring; and (6) Security.

1. **Scalability**: One possible cause of scalability concerns in SDN is the disjointness of the data and control planes [19] and the concentration of control logic in a software-driven controller. The centralized SDN controller receives a large number of events and requests as the network grows in terms of switches, hosts, and other components. This causes an overload of resources, including memory, processing power, and bandwidth. Furthermore, as the network expands in diameter and size, there may be an increase in communication delays between the SDN controller and the network switches, which will affect flow-setup latencies. This could lead to longer failover times as well as congestion in the control and bearer planes [20]. A single physically-centralized SDN controller, however, might not be able to meet the performance requirements (with regard to these metrics) of extensive networks in comparison to small or medium-sized networks, given that throughput—the number of flow requests handled per second—and flow setup latency—the time it takes to respond to flow requests—metrics are frequently used in SDN to evaluate control plane scalability [21]. One way to address some of these scalability issues is to increase the bearer plane's responsibilities in order to lighten the controller's workload. That method's primary flaw is that it requires some changes to the OpenFlow switches' design.

   We think that modeling the control plane in a way that minimizes scalability constraints is a more effective approach. All requests from SDN switches are handled by a single SDN controller in a physically-centralized control model. The latter is expected to develop into a significant performance and scalability bottleneck as the network expands [22]. A physically distributed control model, on the other hand, makes use of several controllers to uphold a logically centralized network view. It is appreciated that this solution manages the controller bottleneck, guaranteeing improved network control plane scalability while lowering control-plane latencies.
2. **Reliability**: Reliability issues are regarded as severe in SDN. Actually, the dependability of the SDN control plane is greatly impacted by the data-to-control plane decoupling. If the principal controller in a centralized SDN-based network fails, the network as a whole might collapse. However, in a physically distributed (but logically centralized) controller architecture, the issue of a single point of failure is lessened through the use of multiple controllers. The OpenFlow standard allows a switch to connect to multiple controllers at the same time,

but it does not specify how a distributed SDN controller architecture should be implemented. Each controller with an active connection to the switch can function as a slave, equal, or master thanks to the OpenFlow option. Based on these OpenFlow roles, which highlight the value of controller replication in establishing a highly available SDN control plane, different fault-tolerant controller architectures have implemented different resilience techniques [23]. In order to improve network availability, controllers also support Multi-Chassis Link Aggregation groups and Virtual Router Redundancy Protocol.

3. **Consistency**: The network may go out of control as a result of consistency issues between the control and bearer planes. For instance, let us say that the packets are routed to unwanted paths or subnetworks, the traffic engineering and access control policies might not work as intended. Further performance deterioration or even a network failure (such as forwarding loops and packet dropping) will be caused by the unpredictable network behaviors. Therefore, ensuring consistency between the control and bearer planes is essential.

   Distributed controllers of SDN platforms encounter significant consistency issues in contrast to physically centralized SDN designs [24,25]. It is clear that to maintain a logically centralized network-wide view which facilitates the expansion of control applications, physically distributed SDN controllers must coordinate with one another and ensure the stability of the network state being scattered across them and kept in their shared data structures. It is challenging to uphold high performance and a convenient level of consistency in software-defined networks that are prone to network partitions. Consistency (C), high availability (A), and partition tolerance (P) are typically not possible for SDN networks to achieve instantaneously, according to the CAP theorem realistic to networks [26].

4. **Interoperability**: To encourage the advancement and uptake of SDN, there are concerns regarding the interoperability of SDN networks. The shortage of open principles for inter-controller communications is a main factor limiting interoperability between SDN controller platforms in today's multi-vendor environments. Apart from standardizing the southbound interface, where OpenFlow is the utmost used standard, as of right now, there isn't any open standard for the East-West and Northbound interfaces that would enable interoperability among OpenFlow implementations. We contend that numerous obstacles, such as the diversity of data models employed by SDN controller vendors, stand in the way of the efficient and quick standardization of the SDN East-Westbound interfaces, notwithstanding the nascent standardization initiatives being supported by SDN organizations [27].

## 2.6 FUTURE SCOPE

SDN technology allows for dynamic traffic management in networks, giving users access to more bandwidth. The removal of the need for specialized hardware is also a cost-effective solution. A simplified perspective of the network is offered. It is believed that SDN is the best solution for meeting the new networking needs. Since SDN is still a relatively new technology, research is ongoing to improve its networking efficiency.

SDN will improve a business's security and productivity in the future without raising operating costs.

Common research-area factors like redundancy, latency, reliability, safety, expenses, and availability need to be talked in order to shape the future of SDN. Consequently, before achieving the best possible set of capabilities, network operators should anticipate that SDN's architecture, interfaces, and applications will undergo multiple revisions

Despite being an open-source technology, SDN has a network monitoring system with centralization features, so you shouldn't worry. SDN will develop into a more secure, fully automated, and responsive technology in the near future.

You might also learn about SDMN in the near future, which will allow software to control the mobile network. Additionally, analysts predicted that the global SDN market would expand at an astounding CAGR of about 19.7% to reach $60.2 billion by 2028.

## REFERENCES

[1] N. Mekondo, H. Kobo, and T. E. Mathonsi, The latest developments in software defined networking: Adoption rate and challenges, *2023 IEEE AFRICON*, 2023.

[2] "Introduction to SDN (Software Defined Networking)", NetworkLessons.com, 2021 [online]. https://networklessons.com/cisco/ccna-routing-switching-icnd2-200-105/introduction-to-sdn-software-defined-networking?

[3] E. Amiri, E. Alizadeh, and M. H. Rezvani, "Controller selection in software defined networks using best-worst milti-criteria decision-making," *Bull. Elect. Eng. Inform.*, vol. 9, pp. 1506–1517. 12 Dec. 2021.

[4] N. Handigol, S. Seetharaman, M. Flajslik, N. McKeown, and R. Johari. Plug-n-serve: Load-balancing web traffic using openflow. *ACM SIGCOMM Demo*, 2009.

[5] R. Wang, D. Butnariu, and J. Rexford. Openflow-based server load balancing gone wild, *Workshop of HotICE*, volume 11, 2011.

[6] A. Nayak, A. Reimers, N. Feamster, and R. Clark. Resonance: Dynamic access control for enterprise networks, *Proceedings of the 1st ACM Workshop on Research on Enterprise Networking*, pp. 11–18, ACM, 2009.

[7] A. Gember, P. Prabhu, Z. Ghadiyali, and A. Akella. Toward software defined middlebox networking, *Proceedings of the 11th ACM Workshop on Hot Topics in Networks*, 2012.
[8] H. Shirayanagi, H. Yamada, and K. Kono. Honeyguide: A VM migration-aware network topology for saving energy consumption in data center networks, *2012 IEEE Symposium on Computers and Communications (ISCC)*, pp. 000460–000467, IEEE, 2012.
[9] https://www.spiceworks.com/tech/networking/articles/what-is-sdn/
[10] B. Heller, S. Seetharaman, P. Mahadevan, Y. Yiakoumis, P. Sharma, S. Banerjee, and N. McKeown. Elastictree: Saving energy in data center networks. *Proceedings of the 7th USENIX Conference on Networked Systems Design and Implementation*, pp. 17–17, USENIX Association, 2010.
[11] F. H. Amir Aldeeb and A. Ali Ahmed. Software defined wide area network SD-WAN: Principles and architecture, *4th International African Conference on Current Studies*, Bani Waleed University, Libya, October 2021.
[12] M. B. A. Lucent, D. T. Bui, L. C. Rakuten, R. Douville. Software-defined LANs for interconnected smart environment, *2015 27th International Teletraffic Congress (ITC 27)*, September 2015.
[13] M. Conti, P. Kaliyar, and C. Lal. Reliable group communication protocol for Internet of Things, *Computer Science, Engineering, Environmental Science*, arXiv.org, 9 April 2019.
[14] W. Mahmood, S. D. Nasir, and I. Waqas. A research survey on software defined networking (SDN), *Proc. of the Ninth Intl. Conf. on Advances in Computing, Control and Networking - ACCN 2019*, 2019.
[15] R. Tulloh, M. Iqbal, and A. B. Aji Pamungkas, "Load balancing implementation strategy for various services in software defined network using ONOS controller", *Int. J. Comput. Digit. Syst.*, vol. 12, pp. 1287–1293, 30 Nov. 2022.
[16] Y. Hu, Z. Li, and L. Yao, "EARS: Intelligence-driven experiential network architecture for automatic routing in software-defined networking", *China Communications*, vol. 17, no. 2, pp. 149–162, 1 Feb. 2020.
[17] S. H. Ali Kazmi, F. Qamar, and B. S. Chowdhry, "Survey on joint paradigm of 5G and SDN emerging mobile technologies: Architecture, security, challenges and research directions", *Wirel. Pers. Commun.*, vol. 130, pp. 2753–2800, 19 April 2023.
[18] M. Jammal, T. Singh, A. Shami, R. Asal, and Y. Li, "Software defined networking: State of the art and research challenges", *Comput. Netw.*, vol. 72, pp. 74–98, Oct. 2014.
[19] M. Karakus and A. Durresi, "A survey: Control plane scalability issues and approaches in software-defined networking (SDN)", *Comput. Netw.*, vol. 112, pp. 279–293, Jan. 2017.
[20] K. Qiu, S. Huang, Q. Xu, J. Zhao, X. Wang, S. Secci, "ParaCon: A parallel control plane for scaling up path computation in SDN", *IEEE Trans. Netw. Service Manag.*, vol. 14, no. 4, pp. 978–990, Dec. 2017.
[21] V. Shamugam, I. Murray, J. A. Leong and A. S. Sidhu, "Software defined networking challenges and future direction: A case study of implementing SDN features on OpenStack private cloud", *IOP Conf. Series: Mater. Sci. Eng.*, vol. 121, 2016.

[22] L. Schiff, S. Schmid and P. Kuznetsov, "In-band synchronization for distributed SDN control planes", *SIGCOMM Comput. Commun. Rev.*, vol. 46, no. 1, pp. 37–43, Jan. 2016.

[23] F. Botelho, T. A. Ribeiro, P. Ferreira, F. M. V. Ramos and A. Bessani, "Design and implementation of a consistent data store for a distributed SDN control plane", *Proc. 12th Eur. Depend. Comput. Conf. (EDCC)*, pp. 169–180, 2016.

[24] K. Lei, G. Lin, M. Zhang, K. Li, Q. Li, X. Jing, P. Wang, "Measuring the consistency between data and control plane in SDN", *IEEE, IEEE/ACM Trans. Netw.*, vol. 31, no. 2, pp. 511–525, April 2023.

[25] F. Bannour, S. Souihi, A. Mellouk, "Distributed SDN control: Survey, taxonomy, and challenges", *IEEE Commun. Surv. Tutor.*, vol. 20, no. 1, pp. 333–354, 2018.

[26] D. Kreutz, F. M. V. Ramos, P. Verissimo, C. E. Rothenberg, S. Azodolmolky, and S. Uhlig, "Software-defined networking: A comprehensive survey," https://arxiv.org/abs/1406.0440, 2021.

[27] Khargharia, B. "OSSDN Drives Interoperability", ONF blog, Sep 7 2016 — describing how SDN enables multi-vendor, multi-domain networks via interoperability.

Chapter 3

# Navigating scalability

## Insights into blockchain enabled SDN environment

*N. Gowthami, G. Ramya, S. Tamilselvi, and S. Chidambaram*

## 3.1 INTRODUCTION

Software-defined networking (SDN) has grown from a niche technology into a key part of today's network setup. In the past, people had to set up and maintain networks by hand using fixed structures. This made networks hard to change and grow. Subsequently, SDN emerged and separated the data plane and control plane. This was a major improvement as it allowed for instantaneous configuration and central management. This change impacted not only the operations of the networks, but most other areas as well. It was used to introduce new concepts such as self-running networks and reliability objectives of virtual networks. They are more flexible; more efficient; and provide enhanced security. This is the fundamental idea of SDN: the technology separates two aspects of a network. It is the control plane that determines to the direction in which the data should flow. The data plane delivers that data to where it has to be delivered. Thus, by separating these up, SDN allows users to view and manage the entire network through one interface, via software to operate it.

But SDN is a significant departure from the conventional theory of network management. SDN also impacts culture and technology, and the benefits, such as centralized control, improved flexibility as well as higher efficiency of the network, among every other, cannot be gainsaid. It can be seen that cultural factors can influence the processes and issues related to the subsequent installation and integration of SDN into present organizational structures.

### 3.1.1 Infrastructure compatibility

Before implementing SDN, it is therefore important that one considers a general assessment of their current network topology. This assessment should first include the assessment of the current infrastructures in regard to the SDN technologies, as well as the invention or identification of legacy systems in the network. This assessment will assist in the identification of other hardware and software that can be used with SDN components.

 DOI: 10.1201/9781003504559-3

### 3.1.2 Skill sets

Before we can ensure that network staff are capable of working with SDN technologies, there is a need to assess the current level of technology understanding among the staff. From this check, we shall be able to determine what they already know regarding SDN. To discover where they need to progress, we have to scrutinize their abilities at their basic level. Next, there should be good training programmes as follows. The team will be ready to face SDN setups and day-to-day tasks if they understand the basics of how to set up SDN, tools used to manage it, and their best practices.

### 3.1.3 Management practices

SDN rests on the organizational and automated way to manage the network; it has shifted the traditional ways of how the network is managed. Thus, to fully unleash the SDN capabilities, organizations are to redesign the current practices. This implies utilizing the automation features in provisioning, monitoring, and policy in SDN instead of the manually configured settings and the decentralized management strategies. By adopting these automated approaches, organizations can perhaps control their networks better, faster, and more flexibly – all of which produces a decrease in practical intricacy and a concurrent increase in effectiveness.

### 3.1.4 Tool integration

To realize the full advantage of SDN in managing the network dynamically, then one has to involve automated tools. These tools assist in enhancing the operations that take place in networks to be smooth and fast and reduce on the mistakes that may occur when they are done by hand. However, to optimize their functioning and bring out the best of their operations, these tools should align with how SDN is created. When it is all done well, the automated systems are aligned with the SDN's centralized management. This leads to network management that is more coherent and functioning well.

### 3.1.5 Communication

The identification of the communication goals requires a clearly defined communication plan in order to gain an understanding of the specifics of SDN, such as benefits and goals, and implications on the different stakeholders. These issues should be answered and explained within this strategy regarding how SDN may improve the network and achieve organizational goals. It is advisable to produce regular updates on the progress during the setup process as it will create interest in everyone. Similarly, providing suitable frameworks for building up support systems is crucial in order to enable the staff to adapt to such changes. Providing information to know more and resolution of issues ensures the workers are equipped to handle the shift for a seamless transition in the process and increased success.

### 3.1.6 Collaboration

SDN implementation requires the integration of efforts from the IT operations, the network specialists, software developers, among others. These two teams should be able to discuss issues and address technological issues in a better way, if they engage themselves in effective communication frequently. Thus, cooperation enables them to communicate their strategies, identify disparities at an early stage, and design successful solutions for the success of the organization. It therefore leads to an enhancement of the network management environment, making it more coordinated. Thus, the conditions under which the network management environment will be more united and effective.

### 3.1.7 Security policies

Since SDN brings about changes in how control and data planes are managed, it is important that security measures that guard against vulnerabilities endemic to controllers and data planes are put in place. These measures are as follows: changing the policies to accommodate the new risks, employing tight security measures, and reviewing the network to counter new risks that may be present. It is a requirement to ensure that the best practices of network security and continuity are achieved and maintained as well, with the goal of making the network environment as safe from a security threat point of view as possible.

### 3.1.8 Threat assessment

As it is evident, there are numerous security risks and dangers in SDN, and this requires a proper risk evaluation. For this reason, the newly configured network has improved in showing faults, and better plans for correction are implemented. Hence, SDN solutions that the vendors offer should be evaluated based on the quality of their support, ability to grow together with the organization, and integration with the currently used systems. This way lays a guarantee that the approaches taken will meet the needs of the organization to offer consistent, enduring support to SDN.

### 3.1.9 Vendor relationships:

The parties need to open the line of communication and bargain with the suppliers in order to get the support and products the business wants. Proper choice of the SDN technology has to meet the technological needs of organizations and their business strategies at the same time, thus being integrated with the current net-based components and applications. This ensures that issues of interfacing are solved promptly and do not disrupt the organization's goals and configuration; the SDN fixes support the goal and configuration as desired by the organization.

### 3.1.10 Monitoring tools

However, to evaluate the success of the deployment of SDN, tools and protocols that enable monitoring of the performance of the network must be developed. Throughput, latency and the use of resources within the SDN system must be measured to know how well the SDN innovative system works. In this way, these performance data should be updated periodically to define areas that could be optimized. From these findings, an organization can enhance the network performance, identify possible challenges early enough, and be guaranteed the intended performance, reliability, and success of the SDN implementation (Figure 3.1).

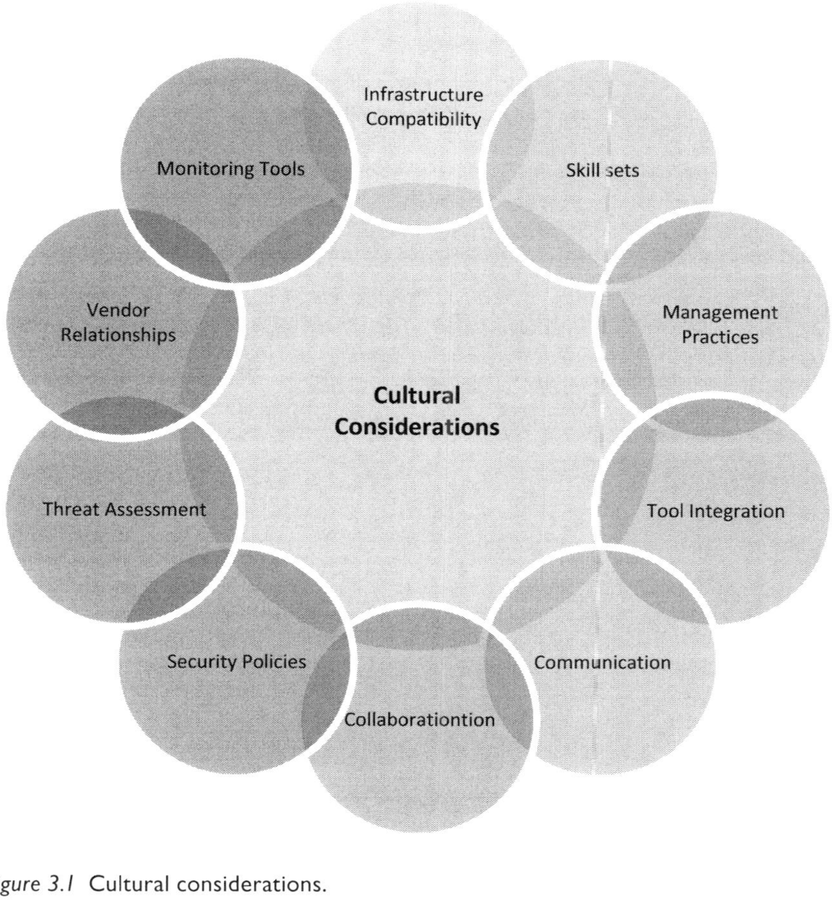


*Figure 3.1* Cultural considerations.

## 3.2 SECURITY AND TRUST ISSUES IN CENTRALIZED NETWORKS

The area of network administration is evolving due to the SDN. It does this by centralized control while physically separating the control plane from the data plane. It has been observed that this setup enhances the functionality of networks and their flexibility as well. But it means new kinds of security concerns at the same time. Solutions are required to prevent the network from being vulnerable and insecure. One of the important security threats in SDN is associated with vulnerabilities of the SDN controller. The controller is actively involved in the management and direction of the traffic in a network, making it a weak link if a person penetrates. Indeed, if a hacker gets into the SDN controller, they may alter the network traffic flow, degrade or even destroy it, or evade security controls. To minimize these dangers, proper controls with regard to access to these resources must be instituted. This helps in limiting the usage and configuration of the controller only to the right people who are supposed to be.

In addition to that, there are some key characteristics of SDN that raise security issues that are listed below.

### 3.2.1 Separation of control and data plane

One of the major concepts which have been adopted in the network management is the establishment of the control plane and data plane. This means it has its own set of security problems that have to be solved to maintain the network secure and running effectively. The control plane is responsible for the management of the network and choices of policies, while the data plane is tasked with handling traffic in the network, thus there has to be safe ways that these two parts of the network can communicate. This is because the control plane is in charge of directing the movements of traffic within the networks, making it extremely sensitive to weak areas of connection.

If an intruder secures a foothold in these communication methods, the latter could access the network's control systems or simply snatch and alter data as it streams through the control/data interface. This could be causing many security issues such as altering rules in a network without prior approval, displaying information that is personal or provoking mayhem on the readily available networks. The general safety of the SDN setup requires that there be secure methods of communication in addition to means of scrambling data. These protect the data flowing through between the control plane and the data plane.

### 3.2.2 Dynamic configuration

This dynamism of changing network setups and policies on the fly is a boon; it is also the bane of SDN. It enables networks to expand and change according to the new conditions and new requirements, and at the same time, it introduces new problems or dangers connected to rapid change. These shifting setups are fast and can be flexible, and thus wrong settings or bad time clashes in the rules involved can easily occur. Such mistakes make it possible for potential attackers to get into the network or disrupt the normal operations. In order to reduce these risks, as well as to keep the network problem-free, it is important that the changes made go through the necessary tests and are known to be benign prior to any further implementation.

### 3.2.3 Open standards and protocols

The basis of SDN is very much embedded in openness between the northbound SDN controller and the southbound SDN device. These standards have benefits regarding important considerations such as interoperability and flexibility. On the other hand, these standards introduce new threats, which are likely to affect the security of the network if not well addressed.

#### *3.2.3.1 Protocol vulnerabilities*

Openness of SDN, which utilizes open standard/protocol, for instance OpenFlow, is one of the security risks of SDN. As designed to allow the SDN controller to interface with multiple network devices, some of these may have inherent flaws in their design that, if not secured, will let attackers exploit them. In this respect, a very serious situation would be protocol implementation defects which grant access to commands governing the network without appropriate permission, which may result in leakage of sensitive information or even control of the physical network. In such a case, measures against these risks must be designed and include encryption, authentication, and SEC periodic tests; security protocols must be immune to any attacks under SDN.

#### *3.2.3.2 Interoperability risks*

Another important advantage of open standards in SDN is interoperability between devices and systems of different vendors. Interoperability also opens doors for security risks within a network. If an implementation of an open standard by one vendor is insecure or has vulnerabilities, then it can bring insecurity to the whole network. This would require careful management and coordination across vendors so that all of the implementations meet stringent security standards. Regular testing, assessment of vendors, and adoption of best practices are much needed to make sure that security vulnerabilities in one part of the network do not taint the whole infrastructure.

### 3.2.4 Southbound interfaces

Southbound interfaces in SDN play a core role in connecting an SDN controller with network devices so that the controller can direct and manage traffic across the network. However, these interfaces also open up specific challenges to the security of the network, which ought to be addressed.

#### *3.2.4.1 Communication with devices*

These southbound interfaces are the ones that carry instructions from the SDN controller to all kinds of different devices within the network. Unless such communication channels are well and securely implemented, they can definitely be a source of weakness. In this case, communications could be monitored by some other entity, or unauthorized commands might be injected. As a result, there would be unauthorized access or manipulation of any one of the network devices. There could be the potential for data breaches, traffic redirection, or even attacks against network services. Any southbound interface must be very well protected, with robust encryption, authentication, and access control mechanisms in place to help avoid such security issues.

#### *3.2.4.2 Protocol exploits*

This is more or less a protocol at the southbound interface, such as OpenFlow, whose precise aim is to facilitate exchanges between an SDN controller and other network devices. The protocols can contain some sweet spots which are not actually disclosed to the public, and an attacker could use these spots. For instance, an attacker might goal at exploiting vulnerabilities in the OpenFlow protocol with an aim of gaining control over network devices, or interrupting the normal network functionality [1]. Such exploits could allow hackers to gain entry into the security line in manners that an attacker could modulate the traffic, modify the configuration, or cause network outages. These risks can be reduced if one is keen to update the protocol implementations and make certain that they are patched, conduct deep security assessments, and highly monitor for the exploits and respond to them when spotted on time.

### 3.2.5 Northbound interfaces: security concerns

The northbound interfaces in SDN are the real interfaces by means of which applications or services interact with the SDN controller. This form of interface introduces flexibility and innovation into the management of networks and, at the same time, exposes some security aspects that must be very well controlled.

#### 3.2.5.1 *Application integration*

In the northbound exposure, the applications expose query, interaction, and modification possibly for the SDN controller for different applications or other services. As it has been mentioned, any strength or weakness of these interfaces can lead to very severe security threats. When such vulnerabilities are exploited, network attackers would be able to gain access directly to the SDN controller in question, this is because the controller is responsible for managing the policy and configuration of the network, which would therefore be impacted. All northbound interfaces need to be implemented and coded to follow access control, strong authentication, and have controllers that must be constantly monitored to detect and counter any intruder accesses to the controller.

#### 3.2.5.2 *API security*

The Application Programming Interfaces (APIs) that are to be exposed for the northbound interfaces remain the most crucial part of the entire integration with the SDN controller, supporting the applications and ensuring smooth interaction. In other effects, such APIs are to be correctly protected through security safeguards so as not to let unauthorized access or ensure integrity in interactions. Unless properly secured, APIs could become a playground for attacks against API abuse, injection attacks, or even data leakage. Nowadays, such risks are decreased by using best security practices, which presently are encryption, authentication tokens, rate limiting, and frequent audits of security. That will ensure that no undesirable application gets access to the SDN controller to allow communication with it.

### 3.2.6 DDoS vulnerabilities

Distributed Denial of Service attacks (DDoS) are one of the most serious security threats within SDN, mainly because of the crucial roles the SDN controller and data plane perform in the functioning of the network [2]. This could logically lead to a situation where the DDoS against these components would make the network unavailable in terms of performance and generally unavailable.

#### *3.2.6.1 Controller and data plane targeting*

The SDN controller is the central management entity of the network and is in charge of making decisions on the flow of traffic and network policies in real time. An SDN-based DDoS attack against the controller floods it with unwanted traffic, overwhelming its processing capacity and hence slowing it down, or even disabling it from managing the network [3]. This can cause a network-wide disruption, loss of control of the running of the network, and probable security breaches. Similarly, the data plane, which is responsible for the actual packet forwarding in accordance with instructions provided by the controller, may also become a target of DDoS attacks. An attacker can flood traffic toward the data plane; this may significantly decrease the performance of the network, resulting in huge latency, considerable packet losses, or even totally bringing down the network.

These threats also demand some effective strategies of DDoS mitigation. These are:

- **Traffic Filtering and Rate Limiting:** This can be based on mechanisms that filter out malicious traffic or rate-limit the processing of traffic to prevent both the controller and the data plane from getting overwhelmed.
- **Redundant Controllers:** The controllers could be considered redundant in case several are running under a distributed architecture; that is, if one goes down due to a DDoS attack or other causes of failure, another can pick up.
- **Scalable Infrastructure:** By scaling both controllers and data-plane resources, an infrastructure can, with the least degradation in performance, help to absorb a DDoS attack.
- **Anomaly Detection:** Continuous monitoring of network traffic for unusual behavior, peaks, or dips in it means that DDoS attacks can be detected early on so that there is enough time to respond and mitigate them.

## 3.3 LIMITATIONS AND VULNERABILITIES OF CENTRALIZED SDN

SDN is the new paradigm in the design and management of computer networks, offering flexibility, control, and programmability never before realized. By decoupling the control plane, SDN allows for the management of the whole network from one central point. Yet, this centralization also introduces a series of limitations and vulnerabilities that should be tackled for the sake of the stability, security, and performance of the deployments regarding SDN.

### 3.3.1 Single point of failure

Probably the best-known vulnerability in centralized SDN deals with a single point of failure, mainly on the part of a centralized controller. The controller is the 'brain' behind an SDN and makes all of the important decisions regarding how the network should be run, from routing and load balancing to policy decisions. In a legacy network, these would be inherent functions of multiple devices, like routers and switches, thus creating redundancy. While in the centralized SDN, if the controller goes down, it can bring an entire network down within minutes. Such dependency on one controller is already introducing considerable risk, especially in environments where network uptime is critical. For example, in financial services, health care, or industrial control systems, even a short network outage may mean losses of millions of dollars, disruption of services, or even human lives at risk. Given the possibility of hardware malfunction, software bugs, or even a focused cyber-attack on the central controller, an SDN could turn out to be a big single point of failure, ultimately leading to far-flung network downtime. In these respects, a failure may mean disaster; therefore, duplicated and backup systems are called for to avoid this type of failure.

### 3.3.2 Scalability challenges

Problems of the scalability of SDN come forth, as with an increasing size and complexity of the network, its centralized nature requires that an SDN controller should have the capabilities of managing and processing an increasingly growing volume of control messages. These include topology changes and control messages from various data points, traffic statistics collected by the network, and failing policies, which the controller will have to process in real time and base the decisions on. With an increase in the number of the connected devices, the load on the controller automatically increases proportionally, and the controller can get congested. Processing of these messages by the controller gets delayed because of the rate of the incoming messages, resulting in an increased latency period, slow rate of response, and eventually, low overall network performance. Geographical limitations keep the scalability issue alive. An example exists in a global enterprise network, where one centralized controller is placed in maybe one country and attached to the devices in other countries or even on other continents. The physical separation between the controller and the devices inevitably brings along latency, which hampers real-time decision-making. Take, for example, an SDN deployment where the controller is physically located in North America and the devices to be managed are in Asia. Latency introduced due to back-and-forth travel time of the control messages causes delays in either policy enforcement or response to other events displayed on the network.

In such cases, some SDN platforms resort to a distributed approach, which deploys a number of controllers throughout the network to cover separate segments or regions. Though this form can offset the load on different controllers, which can mean a decrease in latency, it means opening the door to much greater capability. The guarantee of coordination, synchronization, and consistency across the distributed controllers usually becomes cumbersome; anything that goes wrong results in network inefficiency or conflict, and at times, outages.

### 3.3.3 Security vulnerabilities

This centralized nature of SDN also makes the controller a very juicy target for cyber-attacks since all control decisions are made within the controller. Due to the fact that the SDN controller has an overall view of the network and is in charge of all the control functions, it will be the focus of the attacks, which of course, if an attacker is able to compromise the controller, then it translates to overall control of the entire network, and thus has the capability of not only altering traffic patterns but also tapping into confidential communications, or even attacking other systems connected to it.

This is even more worrying, especially when the integrity and confidentiality of data are paramount, like in the financial institutions, health care providers, and government. An attacker can do a lot worse with the SDN controller—for example, redirecting sensitive data to unauthorized locations, modifying data in flight, or jamming important services through the ability to reroute or drop traffic.

It is also possible that the complexity of the SDN controller software increases the chances of there being many more flaws. Most of the SDN controllers out there are open source, and while that certainly provides flexibility and community-driven innovation, it also means that the exact same codebase is equally available to attackers. This gives them the ability to sniff out and take advantage of any potential holes in the software. Bugs and misconfigurations in the controller code may additionally be utilized by the attackers as an ingress route or method to carry out an operation or activity that may disrupt the network. The risks of this possibility are mitigated by following a frequent update and security patch procedure, but mitigation does not work all the time against elite-class attacks.

Another security implementation can be about the vulnerability of an SDN controller to a DDoS attack. A large number of requests can sink the controller, making it unresponsive and crashing, making the network unavailable. In the case of traditional networks, this would not be such a great issue since the control functions are distributed, meaning that, thereby, one could keep working even if one of them crashed. When the nature of SDN is centralized, a halt of the controller implies that the whole network operation comes to a standstill.

### 3.3.4 Performance bottlenecks

The centralized controller in an SDN deployment has to process a huge volume of information in real time to support informed decisions on how it will manage network traffic. Therefore, such processing requires high computational resources, particularly for large and complex networks. With increasing volumes of control traffic, delays may be caused by the controller in the processing of control requests and sending back the responses, hence causing performance bottlenecks.

This could particularly be a problem in low-latency, responsive communication-oriented applications, like video conferencing, online gaming, or even real-time financial trading. In this case, small delays in control request processing may mean perceptible deterioration of the user's experience or, in some instances, huge financial losses.

Additionally, control communication between the controller and network devices is normally carried out over the control channels. In large networks, control traffic volume occupies a substantial portion of available bandwidth. Insufficient bandwidth for control communication may be one of the reasons for congestion, delayed updates, and hence inefficiency in network management. Finding a balance between real-time control needs and constraints from available bandwidth is one of the critical challenges of centralized SDN deployments.

## 3.4 INTRODUCTION TO BLOCKCHAIN-SDN ECOSYSTEM

In our connected world today, cloud technology plays a key role in driving innovation in the IT industry. This model offers users various services on demand and network access to shared pools of physical resources like computing power and storage. As a result, customers don't need to buy costly hardware to use these services. Instead, they can use everyday devices (like a laptop) with Internet access [4], giving them tools to solve complex problems. Also, cloud computing lets users access resources from anywhere, making virtual teamwork possible. It allows users to boost resources faster than traditional hardware-based systems. Using resources well helps solve the issue of using too much or too little. Cloud computing provides services such as Software as a Service, Platform as a Service, and Infrastructure as a Service. These cloud services are scalable, flexible, and reliable for users when they need them. Blockchain (BC) is a different technology that focuses on security. Because of this, BC technology has caught the eye of the finance sector and other areas where data security is a top concern. Adding secret sharing to BC can enhance security in cloud services by improving data protection. BC technology also helps spot harmful use by making it easier to find suspicious providers.

BC is a decentralized system with a shared ledger that keeps track of transactions from many points in a network [5]. You can't change it, it's safe, and you can see through it so people think it's a great fix for things that need to be secure and trustworthy. Ways to agree, like Proof of Work and Proof of Stake, make sure all points in the network see eye to eye on the BC's state, which is key to keeping it safe. SDN splits the control part from the data part in networking, and this lets you manage the network from one place and change how network resources work on the fly. SDN controllers run network gear like switches and routers by sending orders through the control part, which leads to networks that bend and change more. Putting BC and SDN together helps make networks safer, more open, and tougher. BC offers a record that can't be messed with to keep track of how networks are set up, what security rules are in place, and what moves are made. BC can also spread out SDN control sharing trust between several controllers, which cuts down the chance of everything going wrong because of one weak spot (Figure 3.2).

### 3.4.1 Challenges in scalability and performance

Scalability Issues: Public BCs like Bitcoin and Ethereum struggle to scale due to their consensus mechanisms [6], which restrict transaction speed (e.g., Bitcoin processes around 7 transactions per second). When transactions increase, the network can get clogged. SDN controllers need to manage huge amounts of network data and control many network devices. As the network grows, the controller can turn into a bottleneck affecting overall performance.

The consensus process, transaction validation, and block propagation add latency and computational burden to BC networks. This slows down decision-making and data transmission in an SDN environment. SDN controllers must process flow rules, monitor network state, and react to changes in real time. High traffic volumes can overload the controller, causing delays and lowering network efficiency.

### 3.4.2 Strategies to boost scalability

Sharding splits the BC into smaller sections (shards), each able to process its own transactions. This parallel processing can increase transaction

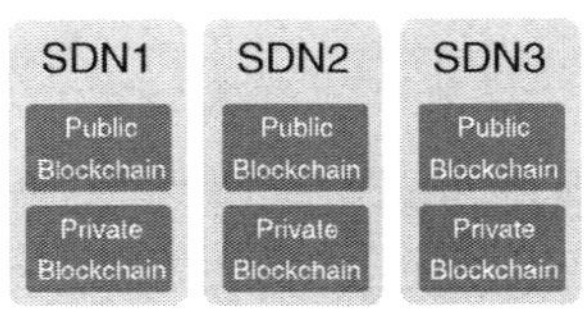


*Figure 3.2* Integration of blockchain and SDN.

throughput. Off-chain solutions like the Lightning Network handle transactions outside the main BC, easing the load on the main chain and improving scalability. Hierarchical SDN Controllers: A hierarchy of controllers can spread out the management load, which makes scaling easier. Controllers at the local level handle specific parts of the network, while a controller at the global level watches over the whole network. SDN can use multiple controllers spread across the network instead of relying on one central controller. This cuts down the chance of bottlenecks.

### 3.4.3 Performance improvements in blockchain-SDN

**Optimized Consensus Algorithms:** Using efficient consensus algorithms (like Proof of Work, Delegated Proof of Stake) cuts down on the computing power and time needed to reach agreement. This boosts overall performance. Consensus methods made for SDN environments can put network performance and reliability first.

**Parallel Processing and Multithreading:** SDN controllers can use parallel processing and multithreading to handle many tasks at once, which cuts down on delays and speeds up response times. In a BC-SDN setup, teams can apply parallel processing to check multiple transactions or deal with several flow rules at the same time.

**Load Balancing and Resource Allocation:** Load balancing tools spread network traffic across different paths or devices, which stops any one controller or device from slowing things down. SDN's ability to shift resources on the fly can make the best use of network resources, making sure key apps get the bandwidth and processing power they need.

**Edge Computing Integration:** Edge computing brings computation and data storage near the data source cutting down delays and bandwidth use. Combining edge computing with BC-SDN can take processing jobs away from the main controller, boosting overall performance.

## 3.5 INTEROPERABILITY CHALLENGES AND STANDARDIZATION INITIATIVES

The integration of BC and SDN has an impact on boosting productivity in smart setups like: smart grid, education, healthcare, industry, transportation, and more. Still, safety remains a big worry for these systems, and we need to work to tackle issues.

Goal and challenges of BC–SDN for safety:

The SDN-based system splits the control plane from the data plane to give networks more flexibility. Yet, the control plane becomes the main target for intruders who can seize it. So, BC-SDN aims to boost the overall security and flexibility of the networking system. BC works well with both private and public environments, and it can help secure P2P talks. To move files in IoT networks, BC offers security, while SDN also helps cut down energy [7] use. For these reasons, we can spot many areas that need the merging of SDN with BC.

Still, some problems exist when combining these two methods, and systems using both face certain issues. First off, bad guys can get into SDN network planes and find out things like host IP or network layout just by doing simple scans. With this info, they can mess with packets and start attacks like ARP or IP spoofing. Also, if someone attacks a host network, they can take over and control messages from the control plane, hijacking it. A man-in-the-middle attack tries to take control of the SDN flow tables and change what's in them.

BC has been suggested as a good way to guard SDN systems, but it's not a cure-all for every SDN security worry, and some issues still need solving. Among possible attacks, DDoS is one of the most harmful since it has an impact on the chain's availability and can let attackers get into linked wallets and exchanges. Also, uninvited users might cause security problems like unauthorized access to wallets and blocking new blocks from joining the chain. What's more, in BC-based setups, deal data sits on public chains, so anyone can see the records. This would create big problems for places where deals need to be secret (like bank transfers). Time and money are two more hurdles when mixing BC with SDN. Since smart systems need to handle many quick-access jobs, less time means better productivity.

### 3.5.1 Confidentiality and privacy in BC–SDN

Security and privacy [8] are big worries when combining BC with SDN. These technologies together can help reach this goal. As we talked about in Section 3.4.1, BC keeps info in blocks after doing cryptography and data hashing [9,10], so privacy is safe, and personal details can stay secret. Since one weak spot can mess up centralized systems, BC's spread-out structure steps in to cut down these weak points. In SDN, data privacy could be at risk when collecting information because raw data comes from different kinds of devices and processing steps can mess with data integrity. Let's look at a real example: setting up security cameras is now super common in smart spaces. Because of this, people can get any privacy because they feel like someone's always watching them, which makes them uncomfortable. BC could help protect privacy so that people who are allowed can see the actual recordings; for everyone else, the system blurs the videos.

Each domain in an SDN uses a private blockchain, while a public BC connects SDN controllers to one another. This approach enhances SDN security. Many studies aim to boost cloud storage performance when combined with BC-SDN. For example, it introduces a distributed BC cloud architecture that relies on software-defined fog nodes [11]. The fog layer's

SDN controllers create a distributed network forming a blockchain, and the cloud layer has its own BC with cloud storage. BC can also help multiple SDN controllers work together smoothly.

Limits of BC-SDN integration link to putting its solutions into practice in the real world. Most studies so far have shown how it works and its perks in test settings, without checking if it's doable. When looking at real-world use, how well current solutions scale up remains a question. For example, if we use BC-SDN to handle and protect lots of devices in an IoT setup, it might slow down a lot or even break down. Also, how well a real system runs could take a big hit depending on which BC system it uses, since the best platforms cost a ton to set up, run, and power; this is still a problem that needs solving in both research and business (like in finance and tech fields).

Another unsolved problem in SDN has to do with synchronization. To be more exact, SDN lacks standard ways to handle synchronization. Also, BC users need to watch out for data privacy, given that BC is visible to everyone who uses it. This might make governments and organizations hesitant to use these technologies when they're dealing with and sharing sensitive information.

It's important to keep in mind that combining BC with SDN can shape their effect on society. Apps that use BC-SDN have better management, openness, and safety. This could rebuild trust between central systems and app users because the data is tracked, checked, and (if needed) open to the public. These unique features offer big chances to impact society, like clear transactions protecting personal info, following rules, and building trust. As we talked about before, this could be used in many areas. From money moves to online voting and health care where important data (money/votes/medical) can be stored safely and shown to the right people when needed, even in emergencies. But while there are lots of good things and chances, it's hard to say for sure if this big change will happen.

Solutions for interoperability play a key role in making communication and transactions smooth between the various networks that have emerged from the growth of BC ecosystems. The strategies below help bridge the gap between different BC networks, spark new ideas, and ensure that BC technology reaches its full potential as a game-changing and unified technology.

#### *3.5.1.1 Cross-chain interoperability*

Cross-chain interoperability is essential to create a unified BC ecosystem. This approach allows different BCs to interact and share data, assets, and smart contracts. It has two main subcategories:

#### 3.5.1.2 Atomic swaps

Atomic swaps allow a transfer of assets from one BC to another without involving a third party. This is an efficient peer-to-peer exchange method that is both secure and very clear.

#### 3.5.1.3 Hashed time-locked contracts (HTLCs)

HTLCs are smart contracts that prevent cross-chain transactions from being fraudulent and provide them with time frames. They ensure that, money gets transferred from one chain to the other, provided some conditions are met and this makes cross-chain transactions safer.

#### 3.5.1.4 Interoperability protocols

BC interfaces can communicate with each other through an interoperability protocol. Some well-known examples of these protocols include:

#### 3.5.1.5 Interledger protocol (ILP)

ILP is an open protocol that functions like the 'internet of value.' The different payment systems can be interconnected through the help of a series of connectors so as to facilitate cross-border transactions.

#### 3.5.1.6 Polkadot

Polkadot builds in parachains that are different blockchains connected through the Relay Chain to the main Polkadot network. In the network, there are several chains of substrates, and the Relay Chain manages consensus, security, and communication between the chains. This setup ensures that different BCs can communicate and exchange information addressing the issue of interoperability.

#### 3.5.1.7 Cosmos

Cosmos architecture is a hub-and-zone approach that designated a few primary blockchains, known as zones, connected to a central hub. This hub enables chains to communicate and exchange information using the Inter-Blockchain Communication.

Interoperability protocols play a great role in enabling communication or interaction between the BC networks. This makes it develop a more connected and productive BC network. These different methods, in a way, create the foundation to link disparate BC platforms and for a more cohesive and integrated progression into the digital world.

#### 3.5.1.8 *Sidechains and state channels*

Sidechains and state channels are also additional techniques for BC connections. Sidechains are other BCs which are linked to the main blockchains in some way. It can communicate with the parent BC, and people employ them to perform a number of tasks without affecting the main BC negatively. The use of sidechains also allows for the ease, speed, and cheap of executing some transactions without straining the main blockchain.

State channels are off-chain fixes that let multiple users make deals or run smart contracts. These actions happen away from the main blockchain, but the end result gets recorded on the blockchain. People use state channels for things like small payments and games.

#### 3.5.1.9 *Middleware and blockchain oracles*

Middleware and Blockchain Oracles act as third-party service provider that helps to interface with the outer world data and other blockchains. They are essential in enabling a way through which BC platforms can interface with the real world and with data. These solutions enable audiences to interact with smart contracts to real-time data and actually connect the BC to the outside world.

#### 3.5.1.10 *Bridge tokens*

Bridge tokens are coins designed to be used for the purpose of interacting between two or more BC ledgers. These tokens are intended to live on several networks, and one can lock a certain amount of them leading to the release of the same amount on the other chain.

## 3.6 CONCLUSION

The integration and operation of BC with software-defined networking is a major historical shift in the practice of managing and protecting networks. A number of innate issues which precipitate the compromising of conventional networking systems like security, transparency, and trust can be solved effectively given the nature of BC technology. BC technologies should be employed in SDN to guarantee that the concessions for policies and configurations cannot be changed and all entrances unauthorized should be blocked. This is specially important nowadays, since threats in cyberspace are more diverse and common that in the past. In addition, by including BC into the system, it is not only possible to realize further decentralization of the system and thereby exclude single points of failure, it enhances overall availability and reliability of the network as well. Nevertheless, there are numerous concerns that have to be resolved to incorporate blockchain into SDN. There exist challenges mostly in the scalability of BC especially in issues to do with handling large numbers of transactions and network topology. On the other hand, the application of BC in real-time transaction has some limitation because of the reaction time that is needed for the formation of consensus in the BC system which needs low-latency communication.

## REFERENCES

1. Priya, A.V., Singh, H.K.: Mitigation of ARP cache poisoning in software-defined networks. In: *Advances in Smart System Technologies*, pp. 85–94. Springer, Singapore (2021).
2. Gadze, J.D., Bamfo-Asante, A.A., Agyemang, J.O., Nunoo-Mensah, H., Opare, K.A.-B.: An investigation into the application of deep learning in the detection and mitigation of DDoS attack on SDN controllers. *Technologies* 9(1), 14 (2021).
3. Thangavel, M., Pavithra, V., Guru Roja, R.: Network manipulation using network scanning in SDN. In: *Artificial Intelligence and Security Challenges in Emerging Networks*, pp. 85–123. IGI Global, Pennsylvania (2019).
4. Aceto, G., Botta, A., Marchetta, P., Persico, V., Pescapè, A.: A comprehensive survey on Internet outages. *J. Netw. Comput. Appl.* 113, 36–63 (2018).
5. Nguyen, G.-T., Kim, K.: A survey about consensus algorithms used in blockchain. *J. Inf. Process. Syst.* 14(1), 101–128 (2018).
6. Altarawneh, A., Sun, F., Brooks, R.R., Hambolu, O., Yu, L., Skjellum, A.: Availability analysis of a permissioned blockchain with a lightweight consensus protocol. *Comput. Secur.* 102, 102098 (2021).
7. Khattak, H.A., Tehreem, K., Almogren, A., Ameer, Z., Din, I.U., Adnan, M.: Dynamic pricing in industrial Internet of Things: Blockchain application for energy management in smart cities. *J. Inf. Secur. Appl.* 55, 102615 (2020).
8. Fitwi, A., Chen, Y., Zhu, S.: A lightweight blockchain-based privacy protection for smart surveillance at the edge. In: *2019 IEEE International Conference on Blockchain (Blockchain)*, pp. 552–555. IEEE (2019).
9. Shao, Z., Zhu, X., Chikuvanyanga, A.M., Zhu, H.: Blockchain-based SDN security guaranteeing algorithm and analysis model. In: *International Conference on Wireless and Satellite Systems*, pp. 348–362. Springer (2019).
10. Yu, Y., Li, Y., Tian, J., Liu, J.: Blockchain-based solutions to security and privacy issues in the Internet of Things. *IEEE Wirel. Commun.* 25(6), 12–18 (2018).
11. Sharma, P.K., Chen, M.-Y., Park, J.H.: A software defined fog node based distributed blockchain cloud architecture for IoT. *IEEE Access* 6, 115–124 (2017).

Chapter 4

# Impact of DDoS attack on cloud enabled networks

*Rakesh Kumar Choudhary, Shriyansh Pandey, and Anu Marwaha*

## 4.1 INTRODUCTION

Cloud computing is a revolution in the IT sector over the past few years. It is basically the process of renting computing resources, instead of owning them over the Internet [1]. Just like using Netflix instead of buying movies. It is the on-demand availability of all kinds of computing resources, be it servers for computing, databases for storage, many networking tools, software, storage, and several other services over the Internet. The unique selling point of Cloud Computing is that it supports *pay-as-u-go* policy, i.e., you need to pay only for those resources that you have used or need. Cloud Computing is beneficial for people from all the sectors, from students to developers, from small organizations to multinational organizations. It's flexible, scalable, cost-effective, fault-tolerant, reliable, and secure, but the security parameter is crucial even after integrating several robust security features as your data resides on a shared platform over the Internet [2].

Cloud security is one of the six pillars of the well-architected framework in cloud computing. It comes with several features to act as a firewall against external threats for data and cloud infrastructure [3]. It involves secure authentication, data encryption, access controls to services and data, and monitoring mechanisms to reduce risks associated with misconfigurations, unauthorized access or privileged access misuse, data breaches, and cyber-attacks [4]. Despite many advancements, data on the cloud remains susceptible to security attacks, that can range from malware injections to causing service disruptions, data loss, and reputational damage. Distributed Denial of Service (DDoS) attacks are one of those severe threats for the cloud environment.

In a DDoS attack, systems are bombarded with continuous and huge traffic from multiple end systems, which results in the network resource being utilized in serving those false requests, making them unavailable for legit users, as shown in Figure 4.1 [5]. This is a major challenge for cloud service providers like Amazon Web Services (AWS), Google Cloud Platform, Microsoft Azure, *etc.*, as well as for users using the services in the cloud, leading to the deployment of reliable DDoS prevention strategies

DOI: 10.1201/9781003504559-4

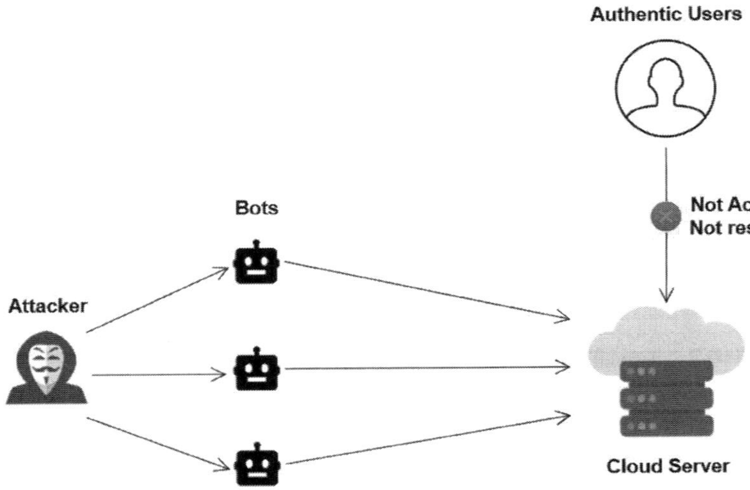


*Figure 4.1* DDoS attack on cloud.

using a multi-layered approach, and scalable infrastructure to ensure seamless service delivery and shield against repercussions.

This research aims to focus on a deeper understanding of the vulnerabilities present in cloud security. Firstly, an overview of the DDoS attacks, their types, and how they are performed has been provided. Certain specific vulnerabilities of cloud services to these attacks were examined, along with the potential consequences for the users. Furthermore, the current methodologies applied to mitigate the DDoS attacks on cloud were assessed. Testing of those preexisting solutions was performed, and constructive criticism of their effectiveness was discussed, along with areas for improvement. Finally, a few prevention measures were proposed that might be helpful in enhancing the cloud security against DDoS attacks.

Furthermore, research has progressed toward cloud security, particularly focusing on DDoS attacks on Platform-as-a-Service (PaaS) frameworks. PaaS environments have been a vital section in cloud computing, which helps developers focus on productivity without thinking about managing the underlying resources.

PaaS refers to the fully furnished platform on cloud that is used for developing and maintaining an application without the responsibility of building or maintaining the underlying infrastructure, i.e., all the necessary resources are taken care of by the cloud service provider, and the user only has to worry about the development and maintenance. It also makes the deployment of applications much more convenient. It is highly scalable, hence making it suitable for different types of organizations according to their usage.

Certain benefits of PaaS include:

1. Reduction in cost
2. Enabling effective DevOps
3. Increased productivity
4. Shorter development cycles
5. Flexible and portable

A few disadvantages of PaaS are:

1. Data is not highly secure.
2. High chances of data mismatch as the data are stored both in local and cloud storage.

However, PaaS has been one of the prime targets for DDoS attacks, which can make services go offline, tamper with data, and cost a hefty sum. Thus, the study further explores and reviews the challenges of securing PaaS setups. The study has successfully examined the existing vulnerabilities and reviewed the existing defences, aiming to take a step further towards cloud security.

The final goal has been to contribute to securing cloud environments and ensuring the safe and uninterrupted use of cloud computing services. This research has taken a step forward in making people aware of taking the necessary steps to make cloud security less vulnerable against one of the most prominent cyberattacks and accelerating further research and development in cloud security. Hence, it makes cloud-based systems more resilient.

## 4.2 LITERATURE REVIEW

This section involves a thorough study of existing insights, highlighting the key findings from various years, as shown in Table 4.1.

The increasing threat of DDoS attacks on the cloud environment and certain ways of mitigation have been described in this paper. It also emphasizes various types of DDoS attacks in cloud computing, distinguishing between external and internal threats across different layers. It highlights attack mechanisms like IP spoofing, SYN flooding, Smurf, buffer overflow, and Ping of Death, alongside corresponding defence measures [1].

The study provides a comprehensive guide for organizations to lessen the adverse effects of DDoS on cloud computing. It highlights the severity of DDoS attacks, some of which are: loss of productivity, reputational damage, and extensive remediation costs. It also suggests some protective measures to reduce their impact on organizational resources and services. It mentions the proactive measures that shall be taken before, during and after the attack [2].

*Table 4.1* Review table

| | *Year* | *Techniques used* | *Parameters* | *Outcomes* |
|---|---|---|---|---|
| [1] | 2013 | SYN cache and cookies in PaaS and Firewall in IaaS | TCP/IP traffic and Detected SYN flooding | Established legitimate connection, increase latency |
| [2] | 2022 | Implement edge defenses and designed high-availability methods | Identified vulnerable services and Block malicious traffic | Robust protections |
| [3] | 2017 | Data Collection | Service/Usage Alterations and Policy Violations | Real-time Analysis Feasibility and Simulating Anomalies |
| [4] | 2015 | Game Theory and 3-Stage Application Layer | Rate Limiting and Request Sequence Analysis | Modern Attack Detection and Congestion control |
| [5] | 2020 | Analysis of DDoS Attacks | Cloud Computing Environments, Vulnerabilities | Strategies for Detecting & Mitigating |
| [6] | 2017 | Service Resizing using processor affinity & TCP tuning | CPU resource allocation, service isolation & fin_timeout | Faster clearing of attack connections & reduced downtime |
| [7] | 2014 | Adaptive management algorithm, LQN Model & Traffic Classification | Cost efficiency metric, Cloud delay center & redirected undesirable traffic | Elasticity and DoS mitigation & Verify legitimacy of users |
| [8] | 2021 | Taxonomy & Literature Review | Cloud features enabling DDoS | Categorization & analysis of defence mechanism |
| [9] | 2018 | Real-time Attack Experiments | Real-time Attack Experiments | Observed Attack Absorption |
| [10] | 2017 | Testing on AWS Cloud, Used LOIC, HULK Tools, Deployed WAF, ELB | Different Attack Strategies & Various Config Settings | Analyzed DDoS Resiliency & Mitigation Methods Evaluated |

We got an idea about a broad survey of Denial of Service (DoS) and Distributed Denial of Service (DDoS) attacks targeting cloud computing, and the various defence mechanisms and evaluation techniques used to assess their effectiveness. It details the different types of attacks, attackers, and defence strategies, including different defences against XML-DoS and HTTP-DoS attacks. It says that despite its scalability and elasticity, cloud computing is vulnerable to a variety of attacks, with DoS and DDoS attacks being particularly prevalent and destructive. While various solutions comprising, the newest technology exist to mitigate these threats, none are foolproof, and system administrators must carefully assess the compromises involved in implementing them [3].

The analytical study makes us aware of the rise of Internet usage and the security threats surrounding it. One such threat is a DDoS attack. It further explains the impacts faced by different industries due to these security risks and their challenges in building defence mechanisms. It also ponders over different types of defence mechanisms, like network-based and application-based approaches. Moreover, it highlights the need for constant research for prevention from DDoS to overcome evolving security threats [4].

Various DDoS methods are examined in this paper, including GET flooding, reflection attacks, and amplification attacks. It outlines the need to secure cloud systems against DDoS attacks by implementing efficient defence mechanisms. It also discusses low-load systems for Wireless Body Area Networks and the vulnerabilities in the cloud-based models. Countering DDoS attacks is vitally dependent on cloud-based monitoring systems and scalable defence mechanisms. Aggregating research findings and deploying VMs can enhance DDoS attack response and security [5].

Various aspects of DDoS attacks and mitigation techniques in cloud computing environments are mentioned by the author. It highlights the differences between regular and extreme DDoS attacks and conducts experiments to study their effects on cloud services. A novel framework for DDoS mitigation is proposed by the authors, which includes an affinity-based victim-service resizing algorithm and a TCP tuning technique to minimize the attack cooling-down period. Performance isolation, freeing up resources for mitigation, and reducing service downtime are the goals to be provided via the proposed techniques. The experimental evaluation demonstrates significant improvements in attack detection time, downtime, and the number of attack requests served, showcasing the effectiveness of the proposed framework in enabling quick DDoS mitigation in cloud environments [6].

An in-depth analysis advises a model-based robust management tools for web applications deployed on the cloud. The objective of the approach is to address two prominent issues: elasticity (scaling resources up or down based on workload) and abatement of DoS attacks. A novel algorithm is suggested by the authors that differentiates between acceptable and undesirable traffic based on a costing metric, which regards both business value and infrastructure cost. A layered queuing network model is adapted by them to precisely anticipate short-term cloud performance, regardless of cloud variability. The approach is evaluated through experiments on Amazon EC2, demonstrating its ability to scale the application whilst palliating DoS attacks and attaining established performance principles. The authors also discuss a limitation of their implementation related to slow reaction time, leading to temporary mitigation of desirable traffic [7].

The author draws attention to prevention strategies against DDoS attacks on cloud computing. The prominence of cloud computing and the adverse impact of DDoS attacks on cloud services are highlighted at the beginning of the discussion. The scenario of DDoS attacks on cloud infrastructure, including the selection of agents (bots) and the victim, as well as the cloud features that enable these attacks, is described by the authors. Several types of DDoS attacks, like SYN flooding, Ping of Death, and User Datagram Protocol (UDP) flood attacks, are discussed. The paper then presents a taxonomy of defence mechanisms, categorized into prevention, detection, and mitigation techniques. Various approaches, including filtering, Challenge Response Protocols, anomaly-based detection, and resource scaling, are explored. Additionally, the authors review proposed solutions from other research papers and provide a comparison of the different defence mechanisms [8].

An innovative approach called "Scale Inside-Out" to quickly palliate the adverse effects of the attack is presented in the paper. It is argued by the authors that during massive DDoS attacks, the victim's system becomes heavily stressed, leading to resource contention and delayed attack absorption, which delays the overall abating process. The proposed approach aims to minimize the absorption delay of the attack by reducing the "Resource Utilization Factor" for each incoming request at that time. This is achieved by renouncing the victim service resources during the downtime, allowing more resources to be dedicated to the palliation service. Experimental results demonstrate that the "Scale Inside-Out" approach can reduce the total downtime after an attack by up to 95% and automatically improve attack detection time [9].

The experiment-based observations of DDoS attacks mitigation process on AWS are primarily shown by the author Waseem Khan in his project report. His paper depicted different types of DDoS attacks, like TCP SYN flooding, UDP flooding, ICMP flooding, HTTP floods, and XML floods. The report outlines two scenarios: testing DDoS attacks directly against servers and testing against servers with Elastic Load Balancer (ELB) and Web Application Firewall (WAF) deployed as part of the AWS mitigation infrastructure. Several open-source attack tools like Low Orbit Ion Cannon (LOIC) and HTTP Unbearable Load King are used to launch attacks in both scenarios. The efficiency of AWS services like ELB and WAF in mitigating different types of DDoS attacks is evaluated through experiments and analysis of results. The report concludes that following AWS best practices and deploying the recommended mitigation infrastructure can significantly reduce the impact of common DDoS attacks [10].

Some unique challenges posed by DDoS attacks in the cloud due to its scalability and reliability features have been displayed by the writers. They divided DDoS defence mechanisms into source-based, network-based, destination-based, and hybrid approaches, analysing their advantages and disadvantages. The paper also ponders over the need for having Software-Defined Networking (SDN) as a mitigation measure for DDoS attacks, as it offers centralized control and programmability for better attack detection and response. It also examines DDoS protective controls like Moving Target Defence, CAPTCHAs, and DNS-based techniques. Overall, the paper presents a thorough survey of DDoS attacks in the cloud, countermeasures, and the enticement of emerging technologies like SDN in mitigating these threats [11].

The study provides various defence mechanisms against DDoS attacks. The different types of DDoS attacks, their detection methods, and some countermeasures are discussed. The study focuses on analysing the effectiveness of different defence strategies, including packet scoring, game theory-based models, application-layer defence mechanisms, and a proposed new combined model. The paper also explores the mathematical analysis of discovered DDoS defence systems and presents a game-theoretical evaluation of the effectiveness of defence mechanisms. The literature review wraps a variety of subjects related to DDoS attacks and their countermeasures, providing intelligence into the latest research and developments in this field [12].

Multiple techniques have been mentioned in the summary table of the literature review to address the adverse impact of DDoS attacks. By implementing SYN cache and cookies in PaaS and implementing firewalls in Infrastructure as a Service, TCP/IP traffic can be managed and SYN flooding can be detected. Moreover, using edge defences and high-availability design methods acts as a helping hand in identifying vulnerable services and providing effective protective solutions against malicious traffic. A continuous assessment of the resilience of the cloud environment to DDoS attacks and the proposed mitigation methods were done to evaluate the performance. Especially focusing on PaaS, the integration of these methods has fortified cloud security to a great extent. PaaS exhibited significant safeguards against SYN flooding attacks, ensuring the integrity of TCP/IP traffic.

## 4.3 METHODOLOGY

The methodology section of this study outlines the approach taken to examine the types and effects of DDoS attacks along with the security measures implemented to mitigate the critical circumstances.

Initially, a broad survey was conducted across multiple platforms, such as Google Scholar, IEEE, ACM, and ResearchGate, exploring various topics ranging from AI to web development to cloud computing.

Among these facets, cloud computing was chosen as it is the need of the hour for various businesses to understand. Cloud computing offers multiple service models, like Software as a Service, PaaS, and Infrastructure as a Service. These service models provide us with six major pillars of the well-architecture framework namely performance efficiency, operational excellence, reliability, security, cost optimization, and sustainability. Still, the security in the cloud is a topic of debate, recognizing the critical nature of this domain, it was decided to channel our efforts towards delving deeper into cloud security.

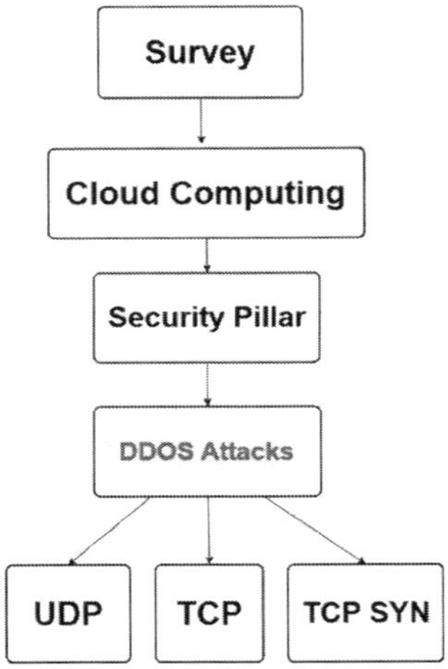


*Figure 4.2* Methodology.

Within the expansive realm of cloud security, a plethora of parameters demanding attention were encountered. With the evaluation of each aspect, we gravitated towards understanding the intricacies of safeguarding cloud environments against malicious attacks. The complete workflow is exhibited in Figure 4.2.

Some of the major risks in security of cloud are: -

- DDoS, where the server doesn't respond to user requests on time.
- Account hijacking, which involves unauthorized access to cloud root or user accounts, often through techniques like phishing or exploitation of weak credentials, causing significant security risks to sensitive data and services.
- Cookie poisoning, in which the malicious content is injected into cookies by the attacker.
- Cloud crypto mining, where attackers use cloud resources to perform crypto mining without the consent of the cloud providers.

Among the array of potential risks, DDoS attacks emerged as a prominent concern due to their disruptive nature and widespread impact on digital infrastructure. By choosing to focus on DDoS attacks, we aimed to address a pressing issue affecting organizations worldwide, seeking to develop insights and solutions to mitigate their detrimental effects. As we delved deeper into the landscape of DDoS attacks, we came across various attack vectors and methodologies used by malicious attackers to disrupt online services. These included TCP Flood, TCP SYN floods, UDP floods, and HTTP floods attacks which are volumetric attacks typically of the POST method but can also be GET method used to engage the maximum resources possible for the server downtime and DNS amplification attacks, which involves exploiting DNS servers to send large amounts of data to targeted victims by spoofing their IP addresses and using DNS queries, resulting in an overload of the system at the victim's end, among others. By selecting DDoS attacks as our area of study, we aimed to gain a comprehensive understanding of these attack types, analyse their mechanisms, and explore effective mitigation strategies. Through meticulous research and analysis, we endeavoured to assist to the progress of powerful defence mechanisms good enough to safeguard cloud-based infrastructures against such malicious activities.

UDP is one of the most preferable DDoS attack vectors used by attackers. This is a type of volumetric attack, where the attacker sends a high volume of UDP packets to target servers, this compels the servers to check for each applications connected with each random port. Then it comprehends that no server is attending to many of these ports, and it then responds the attacker with destination unreachable ICMP error packet. A relatively large amount of packets leaves the server unreachable for the authentic users. Moreover, attackers sometimes spoof their IP addresses to ensure that the ICMP response doesn't reach them, keeping their location anonymous.

TCP flood attacks exploit the susceptibility of the TCP protocol in handling of connection requests and responses. A large number of TCP connection requests to the target server are sent by the attacker. The aim of the flood attack is to use up the hosts or servers by overwhelming them with an extensive number of requests, therefore resulting in failure of response to authentic user requests. It uses other TCP packets along with SYN, like URG (U), ACK (A), PSH (P), RST (R), and FIN (F).

One nefarious DDoS attack is TCP-SYN, SYN short for synchronize. As shown in Figure 4.3, it exploits the three-way handshake method which

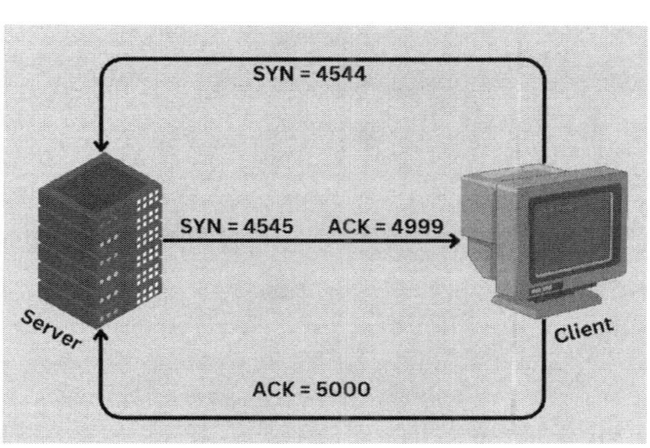


*Figure 4.3* TCP SYN (3-way handshaking).

involves three steps: first a SYN message to the server to demand network is sent by the user, then after acknowledging the request, a SYN-ACK message is replied by the server, finally an acknowledgement (ACK) message is responded by the user and the connection is established. In SYN flood attacks, the attacker sends the SYN message requesting connection and when the server replies with SYN-ACK message, the attacker intentionally avoids sending the last ACK packet rendering the server occupied waiting for it.

### 4.3.1 UDP DDoS attack prevention

1. **Rate Limiting:** This process limits the rate at which the network packets are processed. Implementing lists on the number of UDP packets allowed per second from a single IP address or a subnet helps in mitigation of UDP flood attacks. Cloud providers often offer rate-limiting features in their DDoS protection services.
2. **UDP Protocol Inspection:** Configure the cloud security solutions to inspect UDP packets for malformed headers, invalid payload sizes, or other anomalies that may indicate an attack. Attacks can be reduced or avoided by dropping the packets that fail the inspection.
3. **Blacklisting:** Anonymous sources and sources that are previously known for sending UDP floods can be categorized and maintained in a blacklist of IP addresses. Firewalls and cloud services can then be configured to automatically block the traffic coming from these sources, preventing attack traffic from reaching the target resources to avoid any future repercussions.

### 4.3.2 TCP-SYN DDoS attack prevention

1. **SYN Cookies:** SYN cookies can be implemented to allow servers to track incoming SYN requests without designating servers or other resources until the connection is fully established, as they use cookies as a method to get the previously required data about the connection request. This prevents resource exhaustion during a SYN flood attack.
2. **SYN Proxy:** Using SYN proxy can really be helpful as it serves as a mediator between the client and the server during the TCP handshake process. The proxy handles all the initial SYN requests and forwards only the legitimate connection requests to the actual server by itself, preventing the SYN flood attacks.
3. **Connection Limits:** Cloud security services can be composed to limit the number of concurrent TCP SYN requests received from a single IP address or subnet. This inhibits an attacker from exhausting the server's resources during a SYN flood attack.

4. **Blacklisting:** Sources that have been previously identified for sending TCP SYN floods can be categorized and maintained in a blacklist of IP addresses. Traffic scrubbing services of Cloud can then be used to filtrate malicious SYN traffic before it ambushes the target resources.

### 4.3.3 Mitigation in the cloud

1. **Elastic Scaling:** Elastic scaling is one of the major features provided by cloud computing. This feature automatically scales up or scales down the resources to handle numerous requests at a time. During a SYN flood attack, you can automatically scale up the computing resources to handle the increased load, mitigating the impact of the attack. This is also cost-effective as in case of low usage it scales down the computing resources, hence making it economically feasible as well.
2. **Traffic Scrubbing:** Cloud providers often offer traffic scrubbing services that route incoming traffic through specialized scrubbing centers. These centers can identify and filtrate malicious traffic before it ambushes the target resources, effectively mitigating the attack.
3. **Load Balancing:** routing incoming traffic efficiently is another advantage of using cloud, where all the incoming large number of requests can be routed equally to all the servers to balance the workload, so that the servers can get a sufficient amount of time to mitigate the attack.
4. Some of the major resources of AWS that might lessen the adversity of DDoS are WAF, Shield, CloudFront, ASG, ELB, Route 53.

A major observation that has been carved is that PaaS setups are most prone to UDP attacks, TCP holds second place, and TCP SYN has the best defense mechanism. TCP SYN is more secure than others because it implements three-way handshaking. Moreover, SYN caches also store information about legitimate connections, preventing malicious traffic.

Another major observation is geo-blocking and IP whitelisting. Geo-blocking refers to restricting access to cloud services based on the location of the incoming traffic. By blocking the incoming traffic from the recognized regions for malicious activity, cloud service providers can reduce the probability of attack on the platform to a great extent. Whereas, IP whitelisting acts like marking a green flag for trusted users' IP addresses and marking others as spam connection requests, eventually blocking those malicious ranges.

Moving further, content delivery networks help by distributing content across multiple servers located at various locations, closer to the end user. Since caching and delivering content are done from edge servers, it reduces latency and distributes content more evenly across the network, reducing the chances of DDoS attacks.

## 4.4 CASE STUDY

Google Cloud Armor is a powerful security tool by Google Cloud Platform that provides protection against DDoS attacks and some other application attacks like Cross-Site Scripting (XSS) and SQL Injection (SQLi) [14]. It automatically detects the attack and applies the protective measures as configured manually. On the 1st of June, 2022, around 9:45 a.m. PT, a customer of Google Armor experienced an unparalleled HTTPS DDoS attack, that apexed at 46 million requests per second. This attack was record-breaking, as it turned out to be the largest Layer 7 DDoS attack, around 76% enormous than the previously reported attack. The attack began with more than 10,000 requests per second heckling the HTTP/S Load Balancer. Within 8 minutes, the attack grew to around 100,000 requests per second. However, the Cloud Armor Adaptive Protection detected the attack and issued an alert along with attack signature. The alert also provided a recommended rule to block on the malicious signature. Thereafter, the network security team of the client deployed the recommended rule into their security policy and started to block the attack traffic immediately. On the other hand, the attack enraged and grew to 46 million requests per second. Since the attack traffic was being blocked by the Cloud Armor already, the targeted workload continued to function properly. Ultimately, the attack size started to decrease and completely ended after a span of 69 minutes at 10:54 a.m. It is presumed that the attacker eventually arbitrated that the attack could not have the desired impact over the system. Apart from this high volumetric traffic, the attack also had 5,256 source IPs from 132 countries providing a significant contribution in the attack.

The attack was recognized as the Mēris family of attacks on the basis of the characteristics of the DDoS attack, like geographical distribution and types of unsecured services. Since the customer had already configured the adaptive protection in their Cloud Armor Security policy, it turned out to be a big aid in the time of crisis. The Armor was able to detect the DDoS attack at a very early stage, analyse the traffic, and generate the alert. Further, the customer acted immediately and used rate-limiting feature to tackle the situation as soon as possible [15].

## 4.5 VULNERABILITY

DDoS attacks poses an enormous number of vulnerabilities to the cloud networks worldwide. This is evident from the statistics provided in Figure 4.4 representing the worldwide DDoS attacks in the year 2021. However, some of the major vulnerabilities present in cloud security against DDoS are mentioned below:

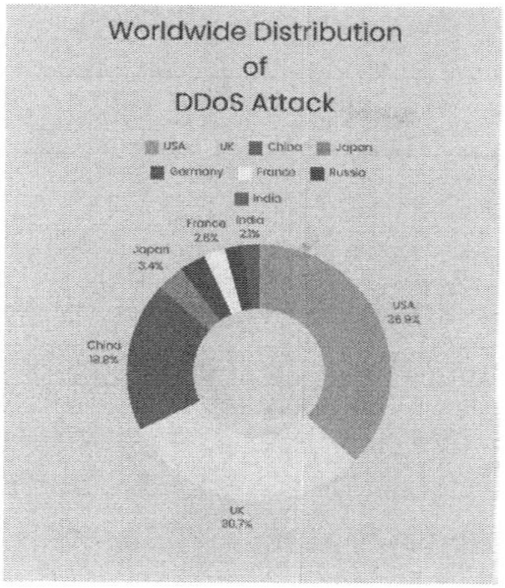


*Figure 4.4* DDoS attacks worldwide 2021 [13].

1. **Insecure APIs and Interfaces**: Multiple studies suggest that APIs and interfaces are often exposed by cloud services. Insufficient authentication, insufficient input validation, and a scarcity of encryption are the major reasons that give the attacker a helping hand in manipulating the services.
2. **Data Privacy**: Cloud computing stores sensitive data in remote data centres managed by third parties, posing challenges to data privacy. For protection against this, users should ensure to keep their data in compliance with industry standards like GDPR, HIPAA, and PCI DSS.
3. **Shared Infrastructure**: Shared infrastructure is also provided by cloud service providers. Multiple users could share the same resources and pay less for this. However, this might be a cost-effective methodology, but it is not a secure one. This enhances the risk of resource contention, where the activity and security of one party could affect the other sharing the same resource. Shared infrastructure also gives space to a wider attack surface, making it highly vulnerable to DDoS attacks, as it is one of the biggest challenges to preventing security breaches.
4. **Network Vulnerabilities**: Cloud computing is heavily dependent on network connectivity for integration and communication between different services. This brings in major vulnerabilities like DNS spoofing and traffic interception. Also, improper configuration in the network setup or weak firewall rules can expose the resources to the attacker, promoting unauthorized access and data breaches.
5. **Lack of Physical Control**: The cloud environment is operated via remote data centres, resulting in a lack of physical control over the infrastructure involved. This is one of the potential concerns where security fails and gives the attacker an upper hand.

These vulnerabilities require a multi-layered solution for cloud security against DDoS attacks.

## 4.6 FUTURE SCOPE

The future scope of research and development in this area is really dense and the scope of advancement lies in multiple directions. As cloud computing is growing exponentially, along with it come major vulnerabilities in cloud security, including DDoS attacks.

Some of the major future scopes are:

1. **ML and DDoS detection using AI**: The current era is called the era of artificial intelligence and machine learning. A good use of AI and ML for the improvement of cloud security against DDoS attacks can be made. Now, even DDoS attacks are becoming more and more difficult to detect; hence, there is an immense need for AI-based detection tools. AI might also be used to read and process network traffic patterns and identify anomalies related to DDoS attacks. Furthermore, the model could be trained to take proactive measures or to take mitigation steps before the disruption of the service.
2. **Blockchain-Based Defence**: Blockchain technology is known for its potential in the field of cybersecurity. Hence, future research can explore the implementation of blockchain-integrated solutions for DDoS defence in cloud environments. Blockchain is resistant to tampering and cyberattacks, so it is beneficial for providing an extra layer of protection and ensuring the availability of services.
3. **Edge Computing**: Since edge computing reduces latency and does the computation closer to the geographical locations of end users, implementing preventive measures closer to the network source can be beneficial to reduce the impact. Furthermore, enabling traffic filtering at edge networks could also be beneficial. This topic might prove to be an interesting area of research.
4. **SDN-Based Mitigation**: Software-defined networks provide a versatile and scalable approach to network management, due to which, it can be used to abate DDoS in a cloud environment. The future scope of an SDN-based mechanism could be to develop a solution using SDN that dynamically reconfigures network traffic flows by rerouting the traffic.
5. **Regenerative Security for the Cloud Environment**: Regenerative security is a new concept that involves self-healing and self-protecting systems capable of recovering from security attacks on their own. This new technique could be integrated with cloud services, enhancing security. However, this needs significant study and research.
6. **Decoy-Based Solution**: Decoy mitigation deploys decoy resources to divert and absorb malicious traffic during the DDoS attack. Further

research could study the effectiveness of implementing this technique in cloud environments. An automatic, scalable tool incorporated with the decoy technique could be developed, as it would be able to handle multiple levels of DDoS attacks because of its scalable nature.

7. **Analysis of Attackers**: a unique approach that will identify and read the attacker's pattern using AI algorithms and other tools. This could vary from identifying their tactics and their usage patterns, as it would be beneficial in identifying and mitigating the attack sooner.

## 4.7 CONCLUSION

On the basis of multiple studies and deep research, we may conclude that DDoS is one of the most hazardous risks to cloud security. Numerous studies have dived deep into the study of different types of attacks, their mechanisms, and their mitigation strategies to minimize their impact. This research stresses the importance of understanding DDoS attack types like UDP floods and TCP SYN floods, along with ways to prevent and mitigate them effectively. Suggestions such as rate limiting, UDP protocol inspection, SYN cookies, and traffic scrubbing are worth mentioning to combat these threats successfully. Additionally, the scalability and flexibility of cloud computing offer methods for mitigation techniques like elastic scaling, load balancing, and the Scale Inside-Out approach intended to reduce downtime while enhancing attack detection and response time. Moreover, the role played by cloud service providers like AWS in providing DDoS mitigation infrastructure, like ELB and WAF, indicates the significance of using cloud resources for security measures.

In summary, as DDoS attacks are evolving in both complexity and magnitude over time, the study offers perspectives on comprehending, averting, and lessening these assaults within cloud computing settings. Employing these strategies enables companies to alleviate the cloud security of their systems. This safeguards uninterrupted service provision to users and secures data from access, safeguarding the company's reputation and trustworthiness. Moreover, the scalability of cloud computing could be made more resilient and safer from DDoS attacks.

## REFERENCES

[1] Darwish M, Ouda A, Capretz LF. Cloud-based DDoS attacks and defenses. In *International Conference on Information Society (i-Society 2013)* (pp. 67–71), June 2013. IEEE.

[2] Cybersecurity and Infrastructure Security Agency, Federal Bureau of Investigation, and Multi-State Information Sharing and Analysis Center. "Understanding and Responding to Distributed Denial-of-Service Attacks." 2022. Available from: https://www.cisa.gov/sites/default/files/publications/understanding-and-responding-to-ddos-attacks_508c.pdf

[3] Bonguet A, Bellaiche M. A Survey of Denial-of-Service and Distributed Denial-of-Service Attacks and Defenses in Cloud Computing. 2017. Available from: https://www.e-helvetica.nb.admin.ch/api/download/urn%3Anbn%3Ach%3Abel-1045487%3Afutureinternet-09-00043.pdf/futureinternet-09-00043.pdf

[4] Sundas A, Badotra S, Bharany S, Almogren A, Tag-ElDin EM, Rehman AU. HealthGuard: An intelligent healthcare system security framework based on machine learning. *Sustainability*. 2022 Sep 22;14(19):11934.

[5] Gupta N, Maashi MS, Tanwar S, Badotra S, Aljebreen M, Bharany S. A comparative study of software defined networking controllers using mininet. *Electronics*. 2022 Aug 29;11(17):2715.

[6] Shuaib M, Badotra S, Khalid MI, Algarni AD, Ullah SS, Bourouis S, Iqbal J, Bharany S, Gundaboina L. A novel optimization for GPU mining using overclocking and undervolting. *Sustainability*. 2022 Jul 16;14(14):8708.

[7] Barna C, Shtern M, Ghanbari H, Litoiu M, Smit M. Model-driven Elasticity and DoS Attack Mitigation in Cloud Environments. In: *Proceedings of the 11th International Conference on Autonomic Computing* (ICAC'14), Philadelphia, PA, USA, June 18–20, 2014. Available from: https://www.usenix.org/conference/icac14/technical-sessions/presentation/barna

[8] Gupta N, Tanwar S, Badotra S, Behal S. Performance Analysis of SDN controller. *International Journal of Performability Engineering*. 2022 Aug 30;18(8):537.

[9] Somani G, Conti M, Gaur MS, Sanghi D, Rajarajan M. Scale Inside-Out: Rapid Mitigation of Cloud DDoS Attacks. *IEEE Transactions on Dependable and Secure Computing*. 2018 Nov/Dec;15(6):959.

[10] Khan WU. DDOS Mitigation Analysis of AWS Cloud Network. Master's Project, University of Victoria; 2017.

[11] Abusaimeh H. Distributed denial of service attacks in cloud computing. *International Journal of Advanced Computer Science and Applications*. 2020; 11(6):163–168.

[12] Sadhu U, Vijaya AKK, Seth K, Riasat MT, Hasan M, Abuzaghleh O. A study on various defense mechanisms against DDoS attacks. *International Journal of Scientific Engineering and Research*. 2015;6(5).

[13] Badotra S, Panda SN. A survey on software defined wide area network. *International Journal of Applied Science and Engineering*. 2020 Mar;17(1):59–73.

[14] Badotra S, Tanwar S, Rana A, Sindhwani N, Kannan R, editors. *Handbook of Augmented and Virtual Reality*. De Gruyter; 2023 Aug 7.

[15] Gundaboina L, Badotra S, Tanwar S. Reducing resource and energy consumption in cryptocurrency mining by using both proof-of-stake algorithm and renewable energy. In: *2022 International Mobile and Embedded Technology Conference (MECON)* (pp. 605–610), March 10, 2022. IEEE.

Chapter 5

# Security revolution for transforming network with blockchain driven SDN

*Medini Gupta, Sarvesh Tanwar, and Monika Nijhawan*

## 5.1 INTRODUCTION

Blockchain is a distributed ledger that keeps track of transactions undertaken in multiple computer nodes in a chronological system by incorporating complicated cryptographic techniques. Verified transactions are stored sequentially inside the blocks with the hash of previous block chained with the header of newly added block [1]. It lays the foundation on a peer-to-peer network where each workstation permits accessibility to computing resources. The decentralised functionality enables transfer of authority from centralised system which is commonly susceptible to a single point of failure to distributed set of networks where every member holds the duplicate copy of the same data [2]. Similar to other technologies, blockchain is also prone to cyber vulnerabilities such as routing and phishing attacks but implementation of strong cryptographic keys can bring a huge difference.

Even when a minimal alteration is performed, the ledger gets turned down by the majority of participants available on the network. The validity of the transaction are ensured through consensus mechanism [3]. It acts as an e-parliament where the primary responsibility of the consensus mechanism is to keep the transactions secure. Minimum 51% of participants on the network must acknowledge the validity of the present state of data. Smart contracts are self-activating agreements where the conditions are predetermined in the code itself [4]. Figure 5.1 displays the workflow of blockchain. With the emergence of smart contracts, involvement of intermediaries is exempted. Let us take a scenario where two entities are going to enter into an agreement, whether for transferring funds or to purchase any asset. In traditional methods, a third party such as a lawyer, agent, or bank was active to look into the agreement and ensure that both the entities satisfy the obligations. On the other hand, the smart contract ratifies the terms of the agreement through set of code.

After the arrival of industrial revolution 3.0, applications of Internet are not just limited to sharing media files but are widely deployed in risk management, thus the demand for imperative security has come [5]. Proper utilisation of data generated from connected gadgets should take place. SDN is

DOI: 10.1201/9781003504559-5

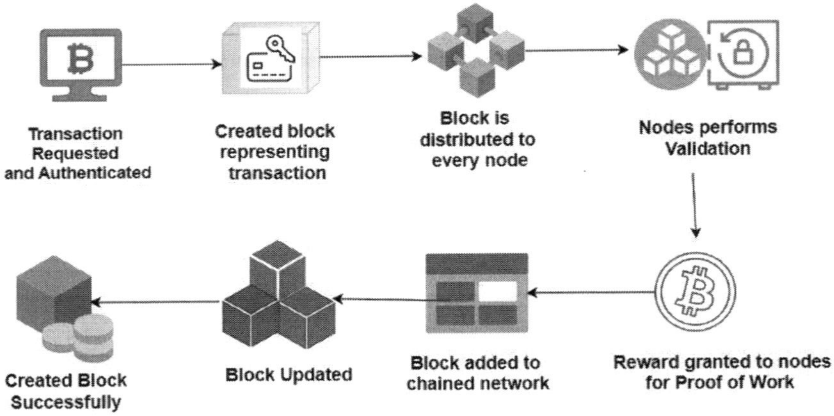


*Figure 5.1* Blockchain workflow.

a sophisticated network architecture widely accepted for monitoring digital security paradigm in automated networks like Internet of Things (IoT) through dedicated software applications. For establishing a fault tolerant system, it is essential to certify that only authorised end users have accessibility for computing resources. This would forbid entry of non-verified entities and lower the recurrence of cyber threats [6]. The architecture of the SDN sets apart the SDN data plane and the SDN control plane which provides centralised management unlike the classic network infrastructure where hardware devices such as switches and routers are banded together in the data plane and control plane for monitoring network congestion.

SDN comprises three integral components, namely control plane, application layer and data plane [7]. Control plane is also commonly referred to as the management plane, that supervises how data packets are carried forward. Routers deploy a number of routing protocols in order to detect network route and these routes are kept in routing tables.

Figure 5.2 reveals the SDN architecture. Alternatively, data planes are also known as packet processing plane or user plane is accountable for transiting network packets from its source to destination that is subject to destination address [8]. The user plane operates at data link layer and network layer of Open Systems Interconnection model whereas control plane functions at network layer only [9]. The application layer is the uppermost layer of Open Systems Interconnection architecture that handles network solution to end users. The applications available with this layer directly interact with control logic plane to effectively monitor the network.

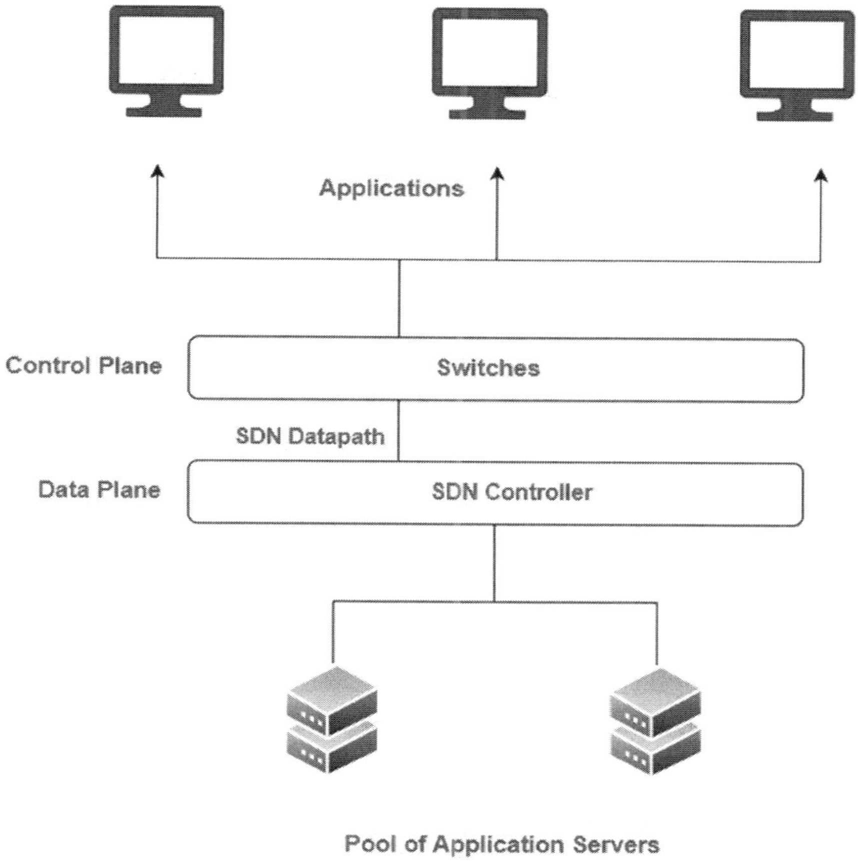


*Figure 5.2* Architecture of SDN.

### 5.1.1 Motivation

Nowadays the maximum number of devices always remains connected to the Internet, the security vulnerabilities that arise from the network devices are also rising. It is estimated that around 80 million devices will be connected to the Internet by 2025 [10]. The surge in network-connected devices will also open the door for cybersecurity threats. Blockchain has come beyond the usage of cryptocurrencies. It has earned tremendous attention due to presence of intrusive combination of numerous frameworks like consensus mechanisms, peer-to-peer network, smart contract and cryptographic techniques [11]. SDN enables an agile and flexible network architecture to assist virtualised servers in a data centre. Pertaining to decentralised nature of blockchain, SDN can make sure that network configurations are auditable and tamper-proof.

### 5.1.2 Research questions

**RQ.1. How can integrity and scalability of SDN architecture be improved by incorporating blockchain based consensus protocols?**

Blockchain consensus protocols such as Proof of Work and Proof of Authority, promises network integrity by laying the foundation for tamper-proof records [12]. SDN can fortify network security infrastructure by deploying cryptographic algorithms for authentication, thus reducing the probability of human errors and increasing the trustworthiness of the transactions undertaken.

**RQ.2. How does blockchain strong set of functionalities make it a compelling opportunity for dynamic resource allocation within SDN while maintaining transparency with minimal overhead?**

Audit trails in blockchain facilitates smooth tracing of resource allotment decision, promoting adherence to guidelines and standards.

**RQ.3. In what ways will blockchain facilitate fault resilience and network transactions at massive scale?** Blockchain is a decentralised ledger that eradicates a single point of failure. Network doesn't get interrupted due to the collapse of a single node [13]. Smart contract can computerize different elements of networking like resources, and dynamically adapt to network system as per demand.

**RQ.4. What are the potential advantages and challenges of running applications in real-time on decentralised server empowered by SDN and blockchain?**

Blockchain with SDN can bring down operational expenses linked with debugging, system administration and digital security [14]. Record verification and consensus protocols of blockchain may lead to latency. It will affect real-time performance and network surveillance in SDN that needs low-latency intercommunication.

### 5.1.3 Paper organization

This chapter consists of various sections. Section 5.1 covered an overview of Blockchain and SDN. In Section 5.2, literature work performed by numerous authors has been discussed and presented in a tabular structure. Section 5.3 involves the vulnerabilities in SDN infrastructure that impact performance and efficiency. Capabilities of blockchain for tackling existing shortcomings of SDN are mentioned in Section 5.4. Then practical deployment for creating and validating blocks is displayed in the next section. Section 5.6 deals with roadblocks in widespread adoption of SDN and blockchain. At the end, we concluded the work by discussing the future scope of these groundbreaking technologies.

## 5.2 LITERATURE REVIEW

In this section, the background work undertaken by various authors in the domain of blockchain and SDN are discussed. Table 5.1 represents potential characteristics.

*Table 5.1* Blockchain and SDN solutions

| *S.No.* | *Proposed solution* | *Description* | *Advantages* | *Challenges* |
|---|---|---|---|---|
| 1 | BlockSD-5GNet [1] | Boosting the security of 5G networks by integrating blockchain with ML bandwidth and Software-Defined Networking (SDN). | Reduced cyber threats through encryption techniques of blockchain.<br>Reliable bandwidth prediction with ML techniques.<br>Deliver optimal performance with minimal resources. | Interoperability issue due to already established network architecture.<br>Centralised nature of SDN can cause single point of failure. |
| 2 | BlockCare [5] | SDN with blockchain for monitoring Internet of Things elements | High scalability due to dynamic scaling of network by SDN.<br>Decentralised control increased robustness of medical system by reducing issue of single point of failure. | Resource-intensive and expensive as it consumes high computing power.<br>Mismanagement of high data traffic leads to network congestion in SDN. |
| 3 | AutAvailChain [6] | Blockchain technology for automatic and tamper-proof data availability | Security is ensured with authorised access permissions and reducing manual operations through automation.<br>Increased the robustness of data retrieval. | With increasing data volumes, scalability issue will occur.<br>Consensus mechanism might cause network latency. Real time data availability will be impacted. |

*(Continued)*

*Table 5.1 (Continued)* Blockchain and SDN solutions

| *S.No.* | *Proposed solution* | *Description* | *Advantages* | *Challenges* |
|---|---|---|---|---|
| 4 | GATE-BC [7] | Blockchain-SDN based Genetic Algorithm-Powered QoS-Aware Cross Network Traffic Engineering | Network performance is optimised with dynamic network traffic monitoring.<br>Transparency in traffic monitoring through blockchain.<br>Genetic algorithm deployed for streamlining traffic engineering. | Genetic algorithm utilises high computational overhead.<br>Possible latency issues with blockchain consensus protocols delays the network traffic. |
| 5 | DistB-SDCloud [8] | A secure infrastructure for cloud computing Industrial Internet of Things (IIoT) with SDN and blockchain | Elevate efficiency and accountability in IIoT with network flexibility through SDN.<br>Mitigates chances of data breaches and data security with decentralised ledger. | Complexity while integrating cloud computing with IIoT. Professional expertise to handle such issues. |

Dipanjali Kundu et al. [1] proposed an architecture "BlockSD-5GNet" to effectively optimise the reliability and security of 5G networks. Authors have explored the possibilities that might arise after intersecting blockchain, ML network function virtualization and SDN. These technologies are capable of handling severe urgencies associated with privacy, performance and security parameters. Although 5G networks deliver a complete package of an efficient network system but it is also susceptible to cyber vulnerabilities. 5G networks encounter enormous threats including secrecy and protection from malicious entities. In the proposed architecture, Blockchain is accountable for enhancing the confidentiality and transparency whereas SDN monitors the network by splitting it into control plane and data plane.

Anichur Rahman et al. [2] introduced "DistB-SDCloud", for fostering the security of cloud computing for Industrial IoT. A sudden call for cloud computing applications has led to SDN being deployed to boost stability, load balancing and durability of cloud-based solution. Scalability and flexibility with integrity is made possible through decentralised behaviour of blockchain. Performance metrics such as latency, throughput, bandwidth and packet assessment, are used to analyse the potency of proposed infrastructure.

Goutam Saha et al. [3] in their review paper mentioned about the security solutions focusing on blockchain for strengthening SDN-based IoT applications. Some of the challenges confronted by IoT and SDN application involves single points of failure, authorization, high resource utilisation, confidentiality and malicious attacks. Blockchain functionalities make it a good choice for conquering the existing drawbacks that persist in SDN and IoT. The reformed architecture of SDN with IoT delivers enhanced resource availability and is suitable for managing higher computational pressure for addressing the challenges of IoT network.

Indukuri Mani Varma et al. [4] in their survey paper have explored the opportunities and considerations that have been raised by converging SDN with Internet of Vehicles. Owing to rapid development in information technology over the past years, vehicular technology interconnects Internet with transportation. The surge in intensity has resulted in complex route management and traffic routing. SDN has been embraced to deliver security and quality at higher level. Efficient resource allocation, upgraded transportation, secure proof programmability and network authentication, are present in SDN infrastructure. Network is susceptible to distributed Denial of Service attacks, poisoning of network topology, etc. In order to combat these adversities, infrastructure focusing on distributed control plane has been incorporated. Despite that, it has encountered various obstacles and major one is reaching consensus on exchange of data among the controller monitoring numerous domains. Therefore, blockchain has been synergies with SDN and Internet of Vehicles for achieving everlasting security. Blockchain features of temper-proof auditing, decentralisation, immutability and consensus protocols make it a promising option for vehicular sector.

Aniruddha Gokhale et al. [5] proposed a solution “BlockCare”, which blends SDN and blockchain for regulating different IoT elements. Patient health-related parameters are controlled by SDN and Blockchain. SDN maintains the scalability of the proposed architecture by catering blockchain virtual functions for mitigating suspicious attacks. Blockchain is responsible for tackling a number of medical issues. Retaining the confidential information of patients with privacy and secrecy is the major challenge. The privilege of consensus protocols and decentralised structure with blockchain has stimulated its quick adoption in clinical trials, supply chain traceability, smooth interoperability, tamper-proof health records, secure identity management.

## 5.3 VULNERABILITIES OF CENTRALISED SDN

Centralised SDN can cause downtime with disruption in network services, which will affect the entire network infrastructure. Questions on robustness and reliability of network are raised when a system faces a single point of failure. This gives malicious entities an open opportunity to hack the system [15]. Central software controllers are more prone to security risk. It can permit unauthorised access. The performance of the system degrades when data traffic increases, leading to scalability issues. The expense for maintaining a secure centralised system with high bandwidth rates for network services is also on higher side.

### 5.3.1 Controller dependency

SDN centralised controller is an indispensable module that is subject to a single point of failure, that will further increase the threat for cyber vulnerabilities. Controller act as the brain of the SDN, which is accountable for managing traffic routing [16]. Any malfunctioning in its workflow would cause network outage. Additionally, dependency on a single controller further leads to scalability issues while deploying at a large scale.

### 5.3.2 Vendor Lock-In

Instead of physical equipment like routers and switches, SDN makes use of software for network monitoring. Vendor lock-in situation occurs when a client heavily puts money in a SDN solution from a particular organization. When client wants to switch to different SDN vendor then it faces flexibility constraints in terms of management interface, proprietary conditions and hardware [17]. Selecting an appropriate SDN solution that is interoperable with both software and hardware from various vendors will increase the diversity.

### 5.3.3 Complexity

Due to dynamic nature of SDN, conventional mechanisms for security turn less effective. Although, agility is achieved when network configurations are performed programmatically, this increases the complexity associated with the enforcement of comprehensive policies [18]. SDN has brought exclusively new protocols that are multifaceted for designing, deploying and managing.

### 5.3.4 Expensive

Shifting to SDN infrastructure demands an initial investment for new software and hardware services [19]. The expense consists of procuring SDN supportable equipment, employing the SDN components that are mandatory and reshaping network infrastructure. With evolving technologies, SDN needs consistent upgradation to assure security with optimal performance. Failing to invest in regular basis might open a path for cyber vulnerabilities, therefore compromising with the performance [20]. For establishing a productive SDN environment, expert IT professionals and network administrators are required. Training price could be substantive that comprises of price related to certifications and courses.

## 5.4 BLOCKCHAIN TO FORTIFY SDN SECURITY

SDN has exhibited substantial potential for amplifying network configurations. Like any other technology, SDN has also come with its shortcomings. These vulnerabilities can be proficiently addressed with blockchain [21]. SDN often faces scalability issues while working with large network sizes. Table 5.2 displays the emerging case studies. The consensus protocols of distributed ledger lays trust among the participants with its immutable and time-stamped authorised transactions without entertaining intermediaries.

### 5.4.1 Efficiency with automation

Various activities of SDN such as allocation of resources, network configuration and provisioning can be automated with blockchain's smart contract.

*Table 5.2* Real-world blockchain and SDN use cases

| *S. no* | *Company* | *Technology* | *Description* |
|---|---|---|---|
| [1] | Microsoft | Blockchain as a Service (BaaS) | Microsoft uses Azure cloud service for facilitating business for building, managing and deploying blockchain. |
| [2] | Amazon | Amazon Managed Blockchain | It uses Ethereum and Hyperledger Fabric framework for establishing scalable blockchain solution. |
| [3] | Google | B4 | B4 is google private Software-Defined Networking that is responsible for inter connecting its data centre with one another. Based on Silicon ASIC's, it handles more network traffic as compared to its public backbone. |
| [4] | AT&T | AT&T Network on Demand | In this Software-Defined Networking solution, businesses are able to dynamically manage network services, in real time while adjusting bandwidth. |

Smart contract is capable of accelerating process, enhancing network operability and reducing error rate that arise due to manual intervention [22]. Financial transactions in the SDN can also be managed through automated settlement that will work depending on the regulations mentioned on the smart contract.

### 5.4.2 Transparent audibility

Blockchain holds a distributed ledger where all the participants have access on the platform. Users can access transactions details such as sender, timestamp, receiver and total amount transmitted on the SDN environment. Blockchain empowers an extensive traceability of transactions undertaken. This will further bring down the possibility of malicious threats and unauthorised access control and will boost confidentiality [23]. Network personnel can smoothly keep track of network updates that are saved on the ledger.

### 5.4.3 Security

Smart contract can regulate access permissions based on prior conditions, making sure that only authenticated users can update the network infrastructure [24]. Blockchain in SDN can empower an authorised and transparent identity management. Network transactions are kept securely with timestamp. Blockchain provides immutable record keeping which facilitates those changes on the network cannot be processed regressively, thus intensifying network security [25]. Transactions performed are encrypted with highly advanced and proven cryptographic algorithms, that protects sensitive data from going to wrong hands.

### 5.4.4 Fault tolerant

In the decentralised environment of ledger, numerous computers (nodes) maintain the duplicate copies of the records [26]. This feature promotes fault tolerant behaviour by reducing the probability of a single point of failure. For enhancing the SDN architecture, blockchain can assist in the distributed control of computing resources. It will lower down the reliance on centralised solution. Multiple SDN platform can be monitored with seamless communication with blockchain standards, which will boost the reliability of the network [27].

## 5.5 PRACTICAL DEMONSTRATION

Following steps were undertaken to create and validate blocks on the blockchain.

To validate the entire block, crypto library "SHA256" was also imported. Figure 5.3 shows the libraries being imported. This deals with generating a crypto hash. A class with its index, hash of previous block and timestamp was created. Based on these properties, hash value was calculated. The initial block on the ledger is known as genesis block. With fixed values, that block is created. Method "getLatestBlock" displays the most recent block on the chained network. Another method named "addBlock", connects previous hash value with the most recent block and generates a new value. At the end, the "isChainValid" function verifies hash value of each block. Figure 5.4 represents that the block created is valid.

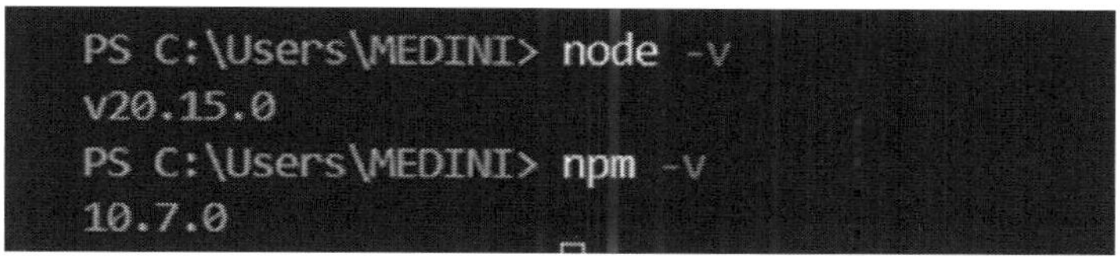


*Figure 5.3* Importing necessary libraries of Node.js.

*Figure 5.4* Blockchain displayed in JSON structure.

## 5.6 ROADBLOCKS IN DEPLOYMENT

SDN architecture works on real-time response and high throughput, that develops challenge due to blockchain's low scalability when transaction volume rises [28]. Dynamic architecture of SDN solution might not fit with blockchain's Proof of Stake and Proof of Importance. Blockchain establishes a decentralised ledger with various nodes performing the verification procedure [29]. This leads to latency challenge. It will affect SDN workflow for achieving flawless network management. Below, we have mentioned the challenges that can arise when SDN and blockchain will shake hands.

### 5.6.1 Complex integration

The challenge of ensuring compatibility with SDN centralised control plane and blockchain decentralised functionality leads to integration complexity [26]. Blockchain architecture is renowned for its peer-to-peer connectivity and distributed consensus protocols that contrast with the SDN, that centrally handles network routing and policies. Blockchain uses its own protocols such as Solidity or Fabric, for reaching on consensus whereas SDN deploys NETCONF or OpenFlow, which possess a communication challenge [9].

### 5.6.2 Poor scalability

With reference to scalability, both SDN and Blockchain need to work on their shortcomings in order to handle increasing data loads. SDN should efficiently monitor network resources that would sync with blockchain consensus protocols like Proof of Importance and Proof of Stake [10]. Low scalability in blockchain directly affects the performance in terms of security and throughput. SDN must go for hybrid infrastructure or distributed controller for tackling escalated network policies for sustaining its performance.

### 5.6.3 Governance

Centralised control of SDN may create a hurdle for tamper-proof ledger if the administrative agreement varies beyond the borders where network controllers are placed. Blockchain solution involves confidential details and regulatory requirement due to its usage in finance and cross-border payment [11]. SDN should be able to manage network traffic proficiently and unify those configurations with auditability, that is necessary for achieving regulatory compliance.

### 5.6.4 Security concern

Cryptographic algorithms form the base for blockchain through which it encrypts confidential information, that makes sure only trusted parties can access the data and this eases the process for reaching on consensus. SDN is more prone to cyber-attacks because of its centralised architecture [12]. Network operations can be compromised if SDN faces any cyber threat. Regular monitoring and vigorous security against malicious vulnerabilities is necessary before converging these technologies together [13].

## 5.7 CONCLUSION AND FUTURE SCOPE

We have delved into how the convergence between Blockchain and SDN can revolutionise computer networks by amplifying security, proficiency and flexibility. Analysed the promising advantages, persistent challenges and diverse real-world business aspects of intersecting these two pioneering technologies. SDN architecture enables network admins to keep track of network activities through lower-level functionality. It is obtained by isolating the control plane, which mainly deals with decisions concerned with traffic transmission, from the data plane that is responsible for traffic rerouting to desired destination. Blockchain refers to a decentralised ledger where every transaction is stored securely in a tamper-proof manner inside a chain of blocks. Consensus protocols are deployed to verify the transactions. Versatile features of blockchain of immutability, decentralisation and timestamped can be maximised to safeguard SDN's control plane. Blockchain can keep track of network updates, that will ease the process of mitigating security vulnerabilities. Regardless of enormous advantages with blockchain and SDN convergence, considerable challenges still need to be addressed. The scalability issue of blockchain persists since its inception while handling a massive amount of network transactions. Executing them on already established infrastructure would raise complications with respect to interoperability and implementation price. Various corporate industries and research initiatives are examining the realistic utilisation of these prominent technologies. Moreover, blockchain is laying the groundwork for establishing a resilient and fortified architecture for SDN that holds the power to monitor network workflow proficiently.

## REFERENCES

[1] Rahman, Anichur, Md Saikat Islam Khan, Antonio Montieri, Md Jahidul Islam, Md Razaul Karim, Mahedi Hasan, Dipanjali Kundu, Mostofa Kamal Nasir, and Antonio Pescapè. "BlockSD-5GNet: Enhancing security of 5G network through blockchain-SDN with ML-based bandwidth prediction." *Transactions on Emerging Telecommunications Technologies* 35, no. 4 (2024): e4965.

[2] Rahman, Anichur, Md Jahidul Islam, Shahab S. Band, Ghulam Muhammad, Kamrul Hasan, and Prayag Tiwari. "Towards a blockchain-SDN-based secure architecture for cloud computing in smart industrial IoT." *Digital Communications and Networks* 9, no. 2 (2023): 411–421.

[3] Indrason, Ngangbam, and Goutam Saha. "Exploring Blockchain-driven security in SDN-based IoT networks." *Journal of Network and Computer Applications* 224 (2024): 103838.

[4] Varma, Indukuri Mani, and Neetesh Kumar. "A comprehensive survey on SDN and blockchain-based secure vehicular networks." *Vehicular Communications* (2023): 100663.

[5] Hakiri, Akram, Aniruddha Gokhale, and Nicolae Tapus. "BlockCare: SDN-enabled blockchain framework for securing decentralized healthcare and precision medicine applications." *Smart Systems for E-Health: WBAN Technologies, Security and Applications* 1 (2021): 81–111.

[6] Camilo, Gustavo F., Gabriel Antonio F. Rebello, Lucas Airam C. de Souza, and Otto Carlos MB Duarte. "AutAvailChain: Automatic and secure data availability through blockchain." In *GLOBECOM 2020-2020 IEEE Global Communications Conference*, pp. 1–6. IEEE, 2020.

[7] Karakus, Murat. "GATE-BC: Genetic algorithm-powered QoS-aware cross-network traffic engineering in blockchain-enabled SDN." *IEEE Access* 12 (2024): 36523–36545.

[8] Rahman, Anichur, Md Jahidul Islam, Shahab S. Band, Ghulam Muhammad, Kamrul Hasan, and Prayag Tiwari. "Towards a blockchain-SDN-based secure architecture for cloud computing in smart industrial IoT." *Digital Communications and Networks* 9, no. 2 (2023): 411–421.

[9] Kakkar, Latika, Deepali Gupta, Sapna Saxena, and Sarvesh Tanwar. "An analysis of integration of internet of things and cloud computing." *Journal of Computational and Theoretical Nanoscience* 16, no. 10 (2019): 4345–4349.

[10] Gupta, Medini, Sarvesh Tanwar, Tarandeep Kaur Bhatia, Sumit Badotra, and Yu-Chen Hu. "A comparative study on blockchain-based distributed public key infrastructure for IoT applications." *Multimedia Tools and Applications* 83, no. 12 (2024): 35471–35496.

[11] Tanwar, Sarvesh, Thomas Paul, Kanwarpreet Singh, Mannat Joshi, and Ajay Rana. "Classification and imapct of cyber threats in India: a review." In *2020 8th International Conference on Reliability, Infocom Technologies and Optimization (Trends and Future Directions) (ICRITO)*, pp. 129–135. IEEE, 2020.

[12] Gupta, Medini, Sarvesh Tanwar, Niranjan Lal, and Ritvik Pawar. "Consensus Algorithms in Blockchain for an Efficient and Secure Network." In *Recent Trends and Best Practices in Industry 4.0*, pp. 309–328. River Publishers, 2023.

[13] Datta, Priyanka, Sarvesh Tanwar, Surya Narayan Panda, and Ajay Rana. "Security and issues of M-Banking: A technical report." In *2020 8th International Conference on Reliability, Infocom Technologies and Optimization (Trends and Future Directions) (ICRITO)*, pp. 1115–1118. IEEE, 2020.

[14] Tselios, Christos, Ilias Politis, and Stavros Kotsopoulos. "Enhancing SDN security for IoT-related deployments through blockchain." In *2017 IEEE Conference on Network Function Virtualization and Software Defined Networks (NFV-SDN)*, pp. 303–308. IEEE, 2017.
[15] Rahman, Anichur, Md Jahidul Islam, Antonio Montieri, Mostofa Kamal Nasir, Md Mahfuz Reza, Shahab S. Band, Antonio Pescape, Mahedi Hasan, Mehdi Sookhak, and Amir Mosavi. "Smartblock-sdn: An optimized blockchain-sdn framework for resource management in iot." *IEEE Access* 9 (2021): 28361–28376.
[16] Hu, J., Reed, M., Thomos, N., AI- Naday, M.F. and Yang, K., 2020. Securing SDN-controlled IoT networks through edge blockchain. *IEEE Internet of Things Journal* 8(4), pp. 2102–2115.
[17] Gao, Jianbin, Kwame Opuni-Boachie Obour Agyekum, Emmanuel Boateng Sifah, Kingsley Nketia Acheampong, Qi Xia, Xiaojiang Du, Mohsen Guizani, and Hu Xia. "A blockchain-SDN-enabled Internet of vehicles environment for fog computing and 5G networks." *IEEE Internet of Things Journal* 7, no. 5 (2019): 4278–4291.
[18] Meng, Weizhi, Wenjuan Li, and Jianying Zhou. "Enhancing the security of blockchain-based software defined networking through trust-based traffic fusion and filtration." *Information Fusion* 70 (2021): 60–71.
[19] Ohri, Pulkit, Subhrendu Guha Neogi, Somenath Sengupta, Daniel Arockiam, and Sunil Kumar Muttoo. "Blockchain-based smart contract architecture for inter-domain SDN controller communication." In *2024 11th International Conference on Reliability, Infocom Technologies and Optimization (Trends and Future Directions) (ICRITO)*, pp. 1–6. IEEE, 2024.
[20] Ghosh, Uttam, Deepak Tosh, Nawab Muhammad Faseeh Qureshi, Ali Kashif Bashir, Al-Sakib Khan Pathan, and Zhaolong Ning. "Cyber-physical systems: prospects, challenges and role in software-defined networking and blockchains." *Future Internet* 14, no. 12 (2022): 382.
[21] Rahman, Anichur, Antonio Montieri, Dipanjali Kundu, Md Razaul Karim, Md Jahidul Islam, Sara Umme, Alfredo Nascita, and Antonio Pescapé. "On the integration of blockchain and sdn: Overview, applications, and future perspectives." *Journal of Network and Systems Management* 30, no. 4 (2022): 73.
[22] Su, Jian, and Mengnan Jiang. "A hybrid entropy and blockchain approach for network security defense in SDN-based IIoT." *Chinese Journal of Electronics* 32, no. 3 (2023): 531–541.
[23] Chaganti, Rajasekhar, Bharat Bhushan, and Vinayakumar Ravi. "A survey on Blockchain solutions in DDoS attacks mitigation: Techniques, open challenges and future directions." *Computer Communications* 197 (2023): 96–112.
[24] Nam Nguyen, Hai, Hai Anh Tran, Scott Fowler, and Sami Souihi. "A survey of Blockchain technologies applied to software-defined networking: Research challenges and solutions." *IET Wireless Sensor Systems* 11, no. 6 (2021): 233–247.

[25] Latif, Sohaib A., Fang B. Xian Wen, Celestine Iwendi, F. Wang Li-Li, Syed Muhammad Mohsin, Zhaoyang Han, and Shahab S. Band. "AI-empowered, blockchain and SDN integrated security architecture for IoT network of cyber physical systems." *Computer Communications* 181 (2022): 274–283.
[26] Jmal, Rihab, Walid Ghabri, Ramzi Guesmi, Badr M. Alshammari, Ahmed S. Alshammari, and Haitham Alsaif. "Distributed blockchain-SDN secure IoT system based on ANN to mitigate DDoS attacks." *Applied Sciences* 13, no. 8 (2023): 4953.
[27] Karakus, Murat, Evrim Guler, and Suleyman Uludag. "Smartcontractchain (SC 2): Cross-ISP QoS traffic management framework with SDN and blockchain." *Peer-to-Peer Networking and Applications* 16, no. 6 (2023): 3003–3020.
[28] Latif, Sohaib A., Fang B. Xian Wen, Celestine Iwendi, F. Wang Li-Li, Syed Muhammad Mohsin, Zhaoyang Han, and Shahab S. Band. "AI-empowered, blockchain and SDN integrated security architecture for IoT network of cyber physical systems." *Computer Communications* 181 (2022): 274–283.
[29] Abou El Houda, Zakaria, Abdelhakim Senhaji Hafid, and Lyes Khoukhi. "Mitfed: A privacy preserving collaborative network attack mitigation framework based on federated learning using sdn and blockchain." *IEEE Transactions on Network Science and Engineering* 10, no. 4 (2023): 1985–2001.

Chapter 6

# Blockchain technology

## Transforming the financial services landscape for banks of the future

*C. Vijai*

### 6.1 INTRODUCTION

The emergence of the blockchain generation is increasingly recognized as a transformative force within the banking industry, offering solutions to the various inefficiencies and security challenges inherent in traditional financial systems. Fundamentally, blockchain serves as a decentralized digital ledger that records transactions across multiple devices in a manner that prevents retroactive alterations. This characteristic ensures transparency, security, and data integrity (Nakamoto, 2008). Within the banking sector, the integration of blockchain technology is poised to revolutionize various aspects of financial services. Traditional banking systems often grapple with slow processing times, high costs, and susceptibility to fraudulent activities and cyber threats. Blockchain effectively tackles these issues by presenting a more secure, efficient, and transparent platform for conducting transactions (Peters & Panayi, 2016). The decentralized structure of blockchain signifies that no single entity wields control over the entire system. Instead, all community contributors own synchronized right of entry to the ledger, ensuring visibility and verification of all transactions by using concerned parties. This transparency diminishes fraud risks and bolsters the credibility of the economic framework (Underwood, 2016). Furthermore, the immutable nature of blockchain ensures that after a transaction is documented, it remains unalterable and unerasable. This attribute is mainly beneficial for audit purposes and regulatory compliance, as it maintains an accurate and tamper-resistant record of all transactions (Yermack, 2017).

Moreover, blockchain facilitates technique automation via smart contracts, which are self-executing agreements with terms immediately encoded into the machine. These intelligent contracts can routinely put into effect and execute agreements whilst specific situations are met, decreasing reliance on intermediaries and expediting transaction processing (Szabo, 1997). Overall, the incorporation of blockchain generation in banking can enhance security, lower expenses, improve performance, and elevate consumer pride. As financial establishments continue to discover and include

DOI: 10.1201/9781003504559-6

blockchain, the panorama of economic services is poised for a great transformation, ushering in more comfortable, obvious, and sustainable banking surroundings.

## 6.2 KEY CAPABILITIES OF BLOCKCHAIN GENERATION

The blockchain era has turned out to be extraordinarily popular due to its talents and capacity to transform several industries, mainly the monetary region. This section explores the vital functions of blockchain technology, providing a comprehensive assessment of its decentralization, transparency, immutability, safety, consensus mechanisms, smart contracts, and the difference between permissioned and permissionless blockchains.

### 6.2.1 Decentralization

One of the critical competencies of the blockchain era is decentralization. Unlike traditional centralized systems wherein facts are managed by a single entity, blockchain operates on an allotted network of nodes. Every node within the network maintains a reproduction of the complete blockchain ledger, making sure that no single entity has completed control over the information (Swan, 2015). This decentralization enhances safety, transparency, and resilience. In a decentralized network, even though one or several nodes fail or are compromised, the community stays operational and the integrity of the data is preserved. Decentralization also reduces the threat of a single factor of failure and makes the system more potent in competition to attacks. For instance, in traditional banking structures, if an essential server is compromised, it could bring about large data breaches and financial losses. In assessment, blockchain's decentralized nature makes it an entire lot extra tough for attackers to alter the information across all nodes simultaneously (Pilkington, 2016).

### 6.2.2 Transparency

Blockchain technology offers high-quality levels of transparency. Every transaction that takes place on the blockchain is recorded in chronological order on a non-public ledger, visible to all people in the community (Nakamoto, 2008). This transparency fosters real trust among clients, as they can independently confirm and audit transactions without counting on the government. Transparency is mainly beneficial in monetary offerings, where belief and obligation are paramount. For instance, in supply chain control, blockchain can track the provenance of products, offering apparent information about the product's journey from its start line to the customer. This can help lessen fraud and boost purchaser confidence (Saberi, Kouhizadeh, Sarkis, & Shen, 2019).

### 6.2.3 Immutability

Immutability is a distinctive key feature of blockchain technology. As quickly as a transaction is recorded on the blockchain, it cannot be altered or deleted. That is completed via cryptographic hashing and the linking of blocks. Each block consists of a hash of the preceding block, developing a series of blocks that can be inherently proof towards alternates (Nakamoto, 2008). The immutability of the blockchain ensures the integrity and accuracy of the statistics. In economic services, this is crucial for maintaining accurate statistics and stopping fraud. For instance, in auditing, blockchain can offer a tamperproof document of economic transactions, improving the reliability of monetary statements and decreasing the risk of fraudulent transactions (Yermack, 2017).

### 6.2.4 Safety

Blockchain employs advanced cryptographic techniques to ease transactions and store statistics in the community. Every transaction on the blockchain is cryptographically signed, ensuring that it can be shown as right (Mougayar, 2016). The decentralized nature of blockchain, furthermore, makes it pretty evidence in the direction of tampering and fraud. Even if one node in the network is compromised, the integrity of the blockchain remains intact because of the allocated consensus mechanism. Furthermore, blockchain networks regularly use public and private keys to authenticate people and conduct transactions. The use of one's cryptographic keys guarantees that quality legal sports can initiate transactions and get proper access to sensitive records (Crosby, Pattanayak, Verma, & Kalyanaraman, 2016). The immoderate degree of protection furnished through blockchain is valuable inside the monetary agency, wherein the protection of touchy records and the prevention of fraud are vital. For example, blockchain may be used to loosen up digital identities, prevent identification robbery, and ensure that the best prison clients can get access to financial offerings (Zheng, Xie, Dai, Chen, & Wang, 2017).

### 6.2.5 Consensus mechanisms

Blockchain technology relies on consensus mechanisms to validate and agree on using a ledger for all nodes in the network. Consensus mechanisms ensure that all transactions added to the blockchain are legitimate and consistent with the network's guidelines (Swan, 2015). There are various types of consensus mechanisms, each with its private blessings and alternate-offs. Proof of Work (PoW): PoW is the most desired consensus mechanism, utilized by Bitcoin and masses of different cryptocurrencies. In PoW, nodes (miners) compete to treat complicated mathematical puzzles, and the first to resolve the puzzle gets to characteristic the subsequent block on the blockchain. This approach requires massive computational energy and strength, making it secure but lacking sizeable (Nakamoto, 2008).

**Proof of Stake (PoS):** PoS is a possibility for PoW that aims to decrease strength intake. In PoS, validators are determined directly to create new blocks based mostly on the number of tokens they keep and are inclined to "stake" as collateral. This approach is a great deal less resource-intensive and provides incentives for individuals to behave as if they have an economic stake within the community (King & Nadal, 2012). Delegated Proof of Stake (DPoS): DPoS is a model of PoS wherein stakeholders pick out a small number of delegates to validate transactions and create new blocks. This mechanism is eco-friendlier and more scalable than PoW and PoS; however, it introduces a level of centralization as electricity is concentered in the hands of some delegates (Larimer, 2014).

**Practical Byzantine Fault Tolerance:** Practical Byzantine Fault Tolerance is completed in permissioned blockchain networks in which individuals are stated and relied upon. Practical Byzantine Fault Tolerance guarantees consensus with the useful resource of requiring a supermajority of nodes to agree on the validity of transactions. This mechanism is green and secure, but it is first-rate and suitable for smaller networks with fewer human beings (Castro & Liskov, 1999). These consensus mechanisms play an essential role in ensuring the protection, reliability, and decentralization of blockchain networks. They prevent malicious actors from gaining management of the community and make sure that all people have a steady view of the blockchain's usage.

### 6.2.6 Clever contracts

Clever contracts are self-executing contracts with the phrases of the agreement written into code right way. These contracts are mechanically positioned under pressure and execute contractual agreements whilst predefined conditions are met (Szabo, 1997). Smart contracts allow programmable and automatic transactions, decreasing the need for intermediaries and streamlining employer techniques. Clever contracts have a large variety of programs in numerous industries, including finance, supply chain management, and real estate. For instance, in the monetary place, clever contracts may be used to automate the settlement of trades, lowering the time and charge related to standard agreement strategies (Kosba, Miller, Shi, Wen, & Papamanthou, 2016). Smart contracts furthermore decorate protection and take shipping as actual by making sure that the terms of the agreement are enforced transparently and robotically. This reduces the risk of disputes and fraud, as all parties may be given the right interest and the agreement can be achieved as agreed (Christidis & Devetsikiotis, 2016).

### 6.2.7 Permissioned vs. permissionless blockchains

Blockchain networks may be labeled as both permissioned and permissionless, depending on their get admission to and governance models (Swan, 2015). Permissionless Blockchains: In a permissionless blockchain, anybody can take part inside the network, validate transactions, and add new blocks to the blockchain. These blockchains are completely decentralized and open to all, presenting excessive levels of transparency and protection. Bitcoin and Ethereum are examples of permissionless blockchains (Nakamoto, 2008; Buterin, 2013). Permissioned Blockchains: Permissioned blockchains limit entry to tremendous participants who are in prison to validate transactions and upload new blocks. These blockchains are usually used by agencies that require extra control over the community and want to conform to regulatory requirements. Permissioned blockchains provide superior scalability and overall performance compared to permissionless blockchains; however, they introduce a diploma of centralization (Androulaki et al., 2018). Permissioned blockchains are typically done in corporation programs, at the side of delivery chain management, finance, and healthcare, in which report privateness and regulatory compliance are crucial. For instance, Hyperledger Fabric is a permissioned blockchain framework designed for business enterprise use, permitting corporations to create and manage non-public blockchain networks (Androulaki et al., 2018).

Blockchain technology's key capabilities of decentralization, transparency, immutability, protection, consensus mechanisms, clever contracts, and the distinction between permissioned and permissionless blockchains make it a powerful tool for remodeling diverse industries. In the monetary quarter, blockchain offers big benefits in terms of safety, overall performance, and transparency, addressing the numerous worrying conditions faced with the beneficial resource of traditional banking systems. As blockchain generation continues to conform, its adoption is expected to expand at some stage in brilliant sectors, riding innovation and reworking company strategies. By leveraging the homes of blockchain, organizations can decorate safety, lessen prices, increase overall performance, and offer higher services to their customers.

## 6.3 ENHANCED SECURITY AND FRAUD REDUCTION IN BLOCKCHAIN TECHNOLOGY

The blockchain era is heralded for its capacity to significantly improve safety and reduce fraud in several sectors, specifically in economic offerings. This section delves into the mechanisms through which blockchain achieves heightened protection and mitigates fraud, supported by instructional references and real-world international examples.

## 6.3.1 Security features of blockchain technology

### 6.3.1.1 Cryptographic security

The blockchain generation employs superior cryptographic strategies to loosen up statistics and transactions. Each transaction on a blockchain is encrypted and connected to the previous transaction using cryptographic hashing (Nakamoto, 2008). This method ensures that any alteration to a block can also require converting all subsequent blocks; which is computationally infeasible, thereby safeguarding the integrity of the facts. Public and private keys are important to blockchain's safety. Users are assigned a private key, which they ought to preserve for their part, and a public key, which they may percent. Transactions are signed with the non-public key and may be shown with the useful resource of every person with the corresponding public key, ensuring the authenticity of the transaction without revealing sensitive statistics (Crosby et al., 2016).

### 6.3.1.2 Decentralization and distributed consensus

One of the core competencies of blockchain is its decentralized nature. In a conventional centralized system, a single entity controls the facts, increasing the hazard of failure. Blockchain, however, operates on a dispersed community of nodes, each keeping a replica of the entire blockchain ledger. This distribution means that even though one node is compromised, the public of the community stays comfortable, and the integrity of the information is preserved (Swan, 2015). Consensus mechanisms, which include PoW and PoS, play an important role in retaining the safety of blockchain networks. These mechanisms make certain that the maximum legitimate transactions are introduced to the blockchain, preventing double spending and wonderful fraud. PoW, for instance, requires miners to solve complicated mathematical puzzles to validate transactions, making it tough for malicious actors to govern the device (Nakamoto, 2008).

### 6.3.1.3 Immutability and tamper-proof records

Blockchain's immutability is a big safety function. Once a transaction is recorded on the blockchain, it cannot be altered or deleted. This immutability is achieved through cryptographic hashing and the linking of blocks. Each block consists of a hash of the previous block, forming a sequence; this is proof of an alternate (Yermack, 2017). The tamperproof nature of blockchain records is mainly precious in monetary offerings and regulatory compliance. It guarantees that transaction histories are accurate and can be audited without the hazard of tampering, thereby enhancing transparency, and agreeing with the tool (Pilkington, 2016).

#### 6.3.1.4 Smart contracts and automated enforcement

Smart contracts are self-executing contracts with terms written into code without delay. They routinely put into effect and execute agreements even as predefined situations are met, lowering the need for intermediaries and minimizing the capability for human errors or manipulation (Szabo, 1997). In the context of safety, clever contracts can automate complex organizational employer techniques and transactions, making sure they are completed precisely as intended. This automation reduces the hazard of fraud and complements the overall performance and reliability of the system (Christidis & Devetsikiotis, 2016).

### 6.3.2 Fraud reduction in blockchain technology

#### 6.3.2.1 Elimination of intermediaries

The blockchain era reduces the need for intermediaries in monetary transactions. Traditional monetary structures depend upon intermediaries, which encompass banks and charge processors, to facilitate transactions, which introduces elements of vulnerability and functionality fraud. By permitting peer-to-peer transactions, blockchain eliminates those intermediaries, decreasing the threat of fraud and enhancing transaction safety (Mougayar, 2016). For instance, in moving border bills, blockchain can streamline the technique with the aid of allowing direct transactions among activities, lowering the time and costs related to well-known banking strategies. This direct interaction minimizes the capability of fraudulent activities and errors, that could arise with a couple of intermediaries (Tapscott & Tapscott, 2016).

#### 6.3.2.2 Real-time monitoring and transparency

Blockchain's transparency and real-time tracking abilities extensively enhance fraud detection and prevention. Every transaction on a blockchain is visible via anybody within the community, imparting an obvious and immutable file that can be audited and set up in actual time (Zheng et al., 2017). This real-time tracking permits the instant detection of suspicious sports activities, permitting quicker responses to fraud. For example, in shipping chain management, blockchain can track the movement of merchandise and affirm their authenticity at each step, lowering the risk of counterfeit products getting into the shipping chain (Saberi et al., 2019).

#### 6.3.2.3 Enhanced identity verification

Blockchain technology enhances identification verification and the usage of digital identities. Traditional identity verification strategies are regularly bulky and at risk of fraud. Blockchain can provide a cozy and immutable document of identities, making sure that humans and entities are as verified as they ought to be (Zyskind, Nathan, & Pentland, 2015). Digital identities on the blockchain are cryptographically relaxed and may be used for a selection of programs, which encompass having access to economic offerings, voting, and more. By imparting a secure and tamper-proof technique of identity verification, blockchain reduces the danger of identity theft and fraud (Shen, Guo, & Yang, 2019).

#### *6.3.2.4 Fraud prevention in financial services*

Blockchain has a huge capability to reduce fraud in monetary services. Traditional financial systems are susceptible to several forms of fraud, which include coin laundering, insider shopping for and promoting, and fraudulent transactions. Blockchain's transparency, immutability, and real-time monitoring capabilities deal with the only vulnerabilities (Gupta, 2017). For instance, blockchain can enhance the protection of securities shopping for and selling by imparting an obvious and immutable file of all transactions. This reduces the hazard of insider shopping for and promoting and ensures that each market member has proper access to identical data, enhancing market integrity (Tapscott & Tapscott, 2016).

## 6.4 STREAMLINING PROCESSES AND REDUCING TRANSACTION COSTS

### 6.4.1 Elimination of intermediaries

Blockchain enables direct peer-to-peer transactions, removing the need for intermediaries, which include banks, charge processors, and clearinghouses. By disposing of the middlemen, blockchain extensively reduces transaction costs and processing times. For example, traditional bypass-border bills can take numerous days and incur excessive expenses because of a couple of middlemen. Blockchain, however, can process the transactions in minutes for a fraction of the fee (Tapscott & Tapscott, 2016). Observe the way Deloitte (2017) determined that blockchain technology has reduced infrastructure prices for eight of the vicinity's ten biggest funding banks with the aid of a median of 30%. These economic and financial savings stem from reduced transaction processing charges, reduced capital requirements, and streamlined operations.

### 6.4.2 Smart contracts

Smart contracts are self-executing contracts with phrases well written into code. They mechanically execute and put into effect the terms of a settlement at the same time as predefined situations are met, reducing the need for manual intervention and minimizing the risk of human errors. This automation results in faster, extra-greener processes and reduces administrative prices (Christidis & Devetsikiotis, 2016). For instance, inside the coverage industry, clever contracts can automate declaration processing. When a policyholder files a declaration, the clever agreement can routinely affirm the declaration information and cause price, significantly lowering processing time and administrative overhead (Marr, 2018).

### 6.4.3 Improved data management

Blockchain offers a single, immutable ledger that every participant in a community can get the right of entry to. This shared ledger gets rid of the want for multiple copies of equal records, lowering garage and protection costs. Additionally, it complements statistics integrity and reduces discrepancies, minimizing the need for reconciliation and audits (Zheng et al., 2017). In supply chain manipulation, blockchain allows real-time monitoring of goods and verification of their provenance. This transparency reduces the risk of fraud and mistakes, main to more green delivery chain operations. For instance, Walmart has carried out a blockchain-based total device to tune the beginning location of its produce, lowering the time it takes to hint meal assets from weeks to seconds (Hyperledger, 2018).

### 6.4.4 Faster transaction settlements

Blockchain technology permits near-on-the-spot agreement of transactions, as antagonistic to standard structures, which may take numerous days. This pace is particularly useful in economic markets, wherein quicker settlement times lessen counterparty chance and loss of capital that would otherwise be tied up in pending transactions (Crosby et al., 2016). For example, the Australian Securities Exchange is converting its gift clearing and agreement tool with a blockchain-based solution. This new tool is expected to provide faster transaction settlements, reduce expenses, and enhance operational efficiencies (ASX, 2017).

### 6.4.5 Decentralized finance (DeFi)

Decentralized Finance (DeFi) leverages the blockchain era to create monetary services and products without traditional intermediaries. DeFi programs, collectively with lending structures and decentralized exchanges, provide lower prices and higher overall performance in contrast to conventional monetary institutions. These systems provide customers with more right of entry to monetary services, in areas with underdeveloped banking infrastructure (Harvey, Ramachandran, & Santoro, 2021). For instance, Compound, a decentralized lending platform, allows clients to lend and borrow cryptocurrencies without the want for a financial enterprise. This peer-to-peer lending version reduces costs and improves the performance of monetary transactions (Compound, 2020).

## 6.5 IMPROVED TRANSPARENCY AND TRACEABILITY WITH BLOCKCHAIN TECHNOLOGY

Blockchain technology is appreciably recognized for its functionality to enhance transparency and traceability throughout numerous industries. By providing a decentralized and immutable ledger, blockchain lets all people in a network view and verify transactions, therefore making sure of more transparency and permitting traceability of property and records. This section delves into how blockchain achieves these advantages, supported by the useful resource of tutorial references and practical examples.

### 6.5.1 Enhancing transparency

#### *6.5.1.1 Immutable ledger*

One of the middle functions of blockchain is its immutable ledger, which ensures that when records are recorded, they cannot be altered or deleted. This immutability is achieved through cryptographic hashing and consensus mechanisms, which make it impossible for any single party to tamper with the information (Nakamoto, 2008). The immutable nature of blockchain enhances transparency by providing a dependable and verifiable report of transactions. For example, within the context of delivery chain management, every transaction from manufacturing to delivery is recorded on the blockchain. This lets all stakeholders, collectively manufacturers, suppliers, and clients, get the right of entry to transparent and unalterable information about the product's adventure. Walmart, for instance, makes use of blockchain to track the beginning and motion of its produce, ensuring transparency in its food delivery chain (Hyperledger, 2018).

#### *6.5.1.2 Decentralization*

Blockchain operates on a decentralized network in which no single entity has control over the entire system. This decentralization distributes power among more than one node, preventing any primary authority from manipulating the facts. This characteristic of blockchain fosters transparency as all participants have the same right to access equal statistics, promoting do not forget and duty (Zheng et al., 2017). Decentralization is mainly beneficial inside the financial region, in which transparency is important for retaining acceptance as true among stakeholders. For example, blockchain- based systems like Ethereum offer apparent and auditable information on transactions, reducing the hazard of fraud and improving recollections in financial operations (Buterin, 2014).

#### *6.5.1.3 Consensus mechanisms*

Blockchain employs consensus mechanisms, such as PoW or PoS, to validate transactions. These mechanisms require agreement from a maximum community of people before a transaction can be introduced to the blockchain. This system ensures that every transaction is transparently tested and agreed upon through the community, further improving transparency (De Angelis et al., 2018). In vote-casting systems, for example, blockchain can offer an apparent and tamper-proof document of votes. Using blockchain, electoral techniques can make sure that votes are efficaciously recorded and counted, thereby developing public self-assurance in the integrity of the election (Shah, 2017).

### 6.5.2 Improving traceability

#### *6.5.2.1 Real-time tracking*

Blockchain allows real-time tracking of property and information, offering a correct and up-to-date document of their reputation. This capability is precious in industries that comprise logistics and supply chain manipulation, in which tracking the motion of products is important (Crosby et al., 2016). For instance, IBM's Food Trust blockchain community lets participants hint at the inspiration and movement of food merchandise through every step of the supply chain. This real-time monitoring facilitates understanding the supply of infection shortages in the event of a foodborne infection outbreak, thereby enhancing food safety and traceability (IBM Food Trust, 2018).

#### *6.5.2.2 Provenance verification*

Blockchain's ability to provide a verifiable file of an asset's journey guarantees the authenticity and provenance of products. This is particularly essential in industries dealing with excessive fee goods, which consist of diamonds, paintings, and comfort gadgets, in which verifying authenticity is important (Pilkington, 2016). Ever Ledger, a blockchain-based platform, tracks the provenance of diamonds from mine to marketplace. By recording every transaction on the blockchain, every ledger ensures that diamonds are struggle-free and authentic, thereby increasing agreement and transparency inside the diamond company (Ever Ledger, 2019).

#### 6.5.2.3 *Automated compliance*

Blockchain can automate compliance with regulatory necessities by presenting an apparent and immutable document of transactions and sports. Smart contracts can be programmed to put compliance into force automatically, decreasing the need for manual oversight and minimizing the chance of noncompliance (Christidis & Devetsikiotis, 2016). In the pharmaceutical enterprise, for example, MediLedger makes use of blockchain to track the movement of pharmaceuticals through the delivery chain. This computerized compliance guarantees that capsules are sourced from legitimate companies and meet regulatory necessities, thereby reducing the threat of counterfeit medicinal tablets (MediLedger, 2019).

### 6.5.3 Case studies and real-world applications

#### 6.5.3.1 *Supply chain transparency*

One fantastic application of blockchain in enhancing transparency and traceability is within the supply chain sector. By presenting an obvious and immutable file of transactions, blockchain permits the monitoring of merchandise from the starting location to the destination. This transparency allows less fraud, beautifies overall performance, and beautifies customer acceptance as true. For example, Provenance, a blockchain-based platform, shall we clients hint on the starting place of products and verify their authenticity? By scanning a QR code on the product, clients can get the right of entry to facts approximately the product's journey, ensuring transparency and agreement (Provenance, 2015).

#### 6.5.3.2 *Financial transparency*

Blockchain's apparent and auditable nature is remodeling the economic vicinity by imparting a clean report of transactions and reducing the hazard of fraud. Blockchain-based total economic systems enable real-time auditing and enhance the transparency of monetary operations, thereby improving regulatory compliance and consideration amongst stakeholders (Yermack, 2017). Ripple, a blockchain-based total fee machine, offers an obvious and efficient solution for cross-border payments. By recording all transactions on the blockchain, Ripple guarantees transparency and traceability, lowering the chance of fraud and enhancing the performance of global payments (Ripple, 2018).

## 6.6 ENHANCED CUSTOMER EXPERIENCE WITH BLOCKCHAIN TECHNOLOGY

The blockchain era is reworking the customer experience for the duration of several industries by way of imparting greater transparency, safety, and efficiency. This section explores how blockchain complements the customer experience through improved, don't forget, streamlined methods, personalized services, and progressive applications. With the aid of academic references and real-world examples, it highlights the profound effect of blockchain on consumer pride and engagement.

### 6.6.1 Trust and transparency

#### *6.6.1.1 Data security and privacy*

Blockchain's decentralized and immutable nature ensures that patron records are comfortable and tamperproof. Using cryptographic strategies, blockchain protects sensitive facts from unauthorized access and cyberattacks, fostering acceptance as authentic among customers (Nakamoto, 2008). In industries like finance and healthcare, where information safety is paramount, blockchain can notably help the purchaser revel in the safety of personal information. For instance, inside the healthcare system, blockchain can securely keep affected individual records, giving patients control over their statistics and making sure that the best prison events can get access to their information. This not only enhances privacy but also builds consideration among healthcare providers (Mettler, 2016).

#### *6.6.1.2 Transparent transactions*

Blockchain's transparency allows clients to tune their transactions in real-time, providing clarity and reducing uncertainties. This transparency is especially beneficial in supply chain manipulation, wherein customers can confirm the authenticity and start of merchandise. By imparting an apparent view of the whole delivery chain, blockchain complements patron belief and satisfaction (Kamath, 2018). For example, every ledger uses blockchain to track the provenance of diamonds, allowing clients to verify the authenticity and moral sourcing of their purchases. This degree of transparency builds agreement with and self-notion in the product, improving the overall client experience (Everledger, 2019).

## 6.6.2 Streamlined processes

### 6.6.2.1 *Faster transactions*

Blockchain permits faster transactions by putting off intermediaries and reducing processing times. In the economic quarter, for example, blockchain can facilitate instantaneous cross-border payments, enhancing the rate and performance of financial offerings. This discount in transaction instances complements client comfort and pleasure (Peters & Panayi, 2016). Ripple, a blockchain-based total charge device, lets in for real-time, low-cost global bills. By using Ripple, economic establishments can methodically go through border transactions in seconds in a matter of days, significantly enhancing the customer experience of international coin transfers (Ripple, 2018).

### 6.6.2.2 *Automated services with smart contracts*

Smart contracts are self-executing contracts with the terms of the settlement immediately written into code. They routinely execute and implement contractual obligations at the same time as predefined situations are met, lowering the need for manual intervention, and minimizing the threat of human errors (Christidis & Devetsikiotis, 2016). This automation streamlines services and enhances efficiency, leading to a better consumer experience. In the coverage business enterprise, for example, smart contracts can automate claim processing. When a claim is filed, the clever agreement can confirm the information and cause charges routinely, lowering processing instances and administrative overhead. This results in faster and additional green service, improving customer delight (Marr, 2018).

## 6.6.3 Personalized services

### 6.6.3.1 *Customer-centric management*

Blockchain permits relaxed and apparent information control, allowing corporations to provide custom-designed services based totally on accurate and up-to-date customer information. By leveraging blockchain, companies can create a single, unified view of customer data, permitting more centered and personalized interactions (Zheng et al., 2017). For example, in a retail location, blockchain can be used to create customized purchasing reports by securely tracking patron opportunities and purchase records. Retailers can use these records to offer personalized guidelines, reductions, and loyalty rewards, enhancing the general purchasing experience (Tapscott & Tapscott, 2016).

#### 6.6.3.2 Decentralized identity management

Blockchain-based, decentralized identity management systems provide clients with control over their digital identities. These structures allow clients to govern and manage their identification records securely, lowering the risk of identity robbery and fraud (Zyskind, Nathan, & Pentland, 2015). Decentralized identity manipulation complements the client's experience by providing a comfortable and accessible manner to verify identity during more than one structure. For example, Civic, a blockchain-based identity management platform, allows customers to create and manage their digital identities securely. By using Civic, customers can confirm their identities quickly and securely, lowering the problem of repeated identity verifications and improving the overall experience (Civic, 2018).

#### 6.6.3.3 Innovative applications

##### 6.6.3.3.1 Loyalty programs

Blockchain generation can revolutionize loyalty applications by using growing interoperable and apparent praise systems. Traditional loyalty programs often suffer from fragmentation and a shortage of transparency, leading to patron dissatisfaction. Blockchain can unify those programs, permitting clients to earn and redeem rewards seamlessly through one-of-a-kind systems and traders (Felin & Lakhani, 2018). Loyal, a blockchain-based loyalty platform, permits the arrival of interoperable loyalty packages. Using Loyal, businesses can offer extra flexible and transparent praise structures, improving customer engagement and pleasure (Loyal, 2019).

##### 6.6.3.3.2 Enhanced customer support

Blockchain can enhance customer support by imparting an obvious and verifiable record of customer interactions. This ensures that each customer support entrepreneur has access to identical data, lowering misunderstandings and improving the performance of aid offerings (Zheng et al., 2017). For example, blockchain-based structures can document all customer support interactions, creating an immutable log that may be accessed by guide dealers. This transparency ensures consistency in customer support and enhances overall consumer enjoyment via the offering of nicely timed and accurate help (Crosby et al., 2016). Blockchain technology offers a massive ability to decorate patron enjoyment with the aid of offering extra transparency, safety, and average overall performance. By making sure records are safe and transparent, streamlining strategies through quicker transactions and clever contracts, allowing custom-designed services, and helping modern packages which include loyalty packages and better customer support, blockchain transforms how companies engage with clients. Despite demanding situations, which include privacy concerns and integration complexities, the continuing improvements within the blockchain era maintain to cope with these issues, paving the manner for its broader adoption in enhancing consumer enjoyment.

## 6.7 REGULATORY COMPLIANCE AND REPORTING WITH BLOCKCHAIN TECHNOLOGY

Blockchain generation affords transformation functionality for regulatory compliance and reporting by enhancing transparency, accuracy, and overall performance. Its decentralized and immutable ledger device guarantees that each transaction is securely recorded and its effects are verifiable, this is critical for preserving regulatory requirements and facilitating reporting procedures. This segment explores how blockchain addresses key stressful conditions in regulatory compliance and reporting, supported by the usage of educational references and practical examples.

### 6.7.1 Enhancing transparency and accuracy

#### *6.7.1.1 Immutable records*

Blockchain era provides an immutable file of all transactions, which guarantees that facts cannot be altered or deleted as quickly as they are recorded. This function is particularly valuable for regulatory compliance, because it ensures the integrity and accuracy of facts, making them dependable for audits and regulatory critiques (Nakamoto, 2008).

For instance, inside the monetary organization, blockchain can make sure that transaction facts are accurate and tamperproof, lowering the risk of fraud and errors. This immutability allows financial institutions to conform with policies that encompass the Sarbanes-Oxley Act, which calls for stringent recordkeeping and accuracy in monetary reporting (Peters & Panayi, 2016).

#### *6.7.1.2 Real-time auditing*

Blockchain allows real-time auditing through the availability of continuous get entry to transaction records. This capability allows regulators and auditors to show compliance in actual time, rather than relying on periodic audits. Real-time auditing improves the overall performance of compliance techniques and decreases the chance of regulatory breaches going undetected (Yermack, 2017). For example, companies can use blockchain to maintain the real-time facts of financial transactions, which may be done right away by auditors. This non-prevent right of access to updated statistics enhances the accuracy and timeliness of audits, ensuring compliance with financial rules (Dai & Vasarhelyi, 2017).

## 6.7.2 Streamlining compliance processes

### 6.7.2.1 *Smart contracts for automated compliance*

Smart contracts are self-executing contracts with the terms of the settlement written into code without delay. These contracts routinely enforce regulatory necessities and enterprise regulations, decreasing the need for guide intervention and minimizing the chance of noncompliance (Christidis & Devetsikiotis, 2016). In the coverage enterprise, for instance, smart contracts can automate compliance exams for policy issuance and claim processing. By embedding regulatory requirements into smart contracts, insurers can make certain that each transaction complies with relevant legal guidelines and regulations, thereby lowering the govt burden and enhancing compliance with universal overall performance (Marr, 2018).

### 6.7.2.2 *Efficient reporting*

Blockchain facilitates green reporting imparting a single delivery of truth for all transactions. This unified ledger guarantees that every stakeholder has the right get admission to regular and accurate facts, simplifying the reporting device and reducing discrepancies. Blockchain's transparency moreover could allow regulators to get the right of entry to important data without requiring massive reporting from organizations (Tapscott & Tapscott, 2016). For instance, in the pharmaceutical business enterprise, blockchain may be used to track the production and distribution of drugs, making ensure compliance with regulatory requirements for traceability and reporting. The U.S. Food and Drug Administration (FDA) has explored blockchain for enhancing the traceability of prescription drugs, aiming to enhance regulatory compliance, and reporting overall performance (FDA, 2019).

## 6.7.3 Facilitating regulatory oversight

### 6.7.3.1 *Enhanced traceability*

Blockchain's potential to provide an obvious and immutable document of transactions enhances traceability, which is crucial for regulatory oversight. Regulators can hint the information about an asset or transaction from its beginning to its present-day United States, ensure compliance with guidelines, and pick out any discrepancies or violations (Zheng et al., 2017). In delivery chain control, for example, blockchain can hint at the motion of products from manufacturing to move, ensuring compliance with regulatory requirements for pride and protection. This more potent traceability allows regulators to display the supply chain greater successfully and ensures that groups adhere to applicable necessities (Kamath, 2018).

#### 6.7.3.2 *Secure data sharing*

Blockchain allows comfy and obvious fact-sharing among more than one party, which is essential for regulatory compliance in industries that require collaboration amongst precise entities. By using blockchain, companies can securely proportion facts with regulators, auditors, and one-of-a-kind stakeholders, making sure that everyone has the proper right of entry to correct and update records (Zyskind, Nathan, & Pentland, 2015). For example, inside the financial context, blockchain can facilitate relaxed record sharing among banks and regulators, improving the transparency and overall performance of regulatory oversight. This ease of sharing of statistics enables compliance with rules, which include the Anti-Money Laundering and Know Your Customer necessities (Felin & Lakhani, 2018).

### 6.7.4 Case studies and applications

#### 6.7.4.1 *Financial services*

In the monetary services agency, blockchain is being used to enhance regulatory compliance and reporting. Companies like J.P. Morgan and Goldman Sachs are exploring blockchain for clearing and agreement tactics, aiming to lessen operational risks and ensure compliance with regulatory requirements (Peters & Panayi, 2016). The Depository Trust & Clearing Corporation (DTCC), which methods most securities transactions in the U.S., is developing a blockchain-based platform to improve publish exchange processing. This platform aims to enhance transparency, lessen operational expenses, and ensure regulatory compliance by supplying a single, immutable document of all transactions (DTCC, 2017).

#### 6.7.4.2 *Healthcare*

In healthcare, blockchain can enhance regulatory compliance by presenting a comfortable and obvious document of patient facts and clinical transactions. The use of blockchain for virtual health records (EHRs) ensures that affected character information is correct, cozy, and available simplest to authorized activities, complying with regulations along with the Health Insurance Portability and Accountability Act (HIPAA) (Mettler, 2016). For example, Medi Bloc is a blockchain-based platform that provides a comfortable and obvious way to manage affected personal information, ensuring compliance with regulatory necessities for data privacy and protection. With the use of MediBloc, healthcare providers can securely share patient statistics, improving the performance and accuracy of clinical records (MediBloc, 2018).

Blockchain generation gives big benefits to regulatory compliance and reporting by improving transparency, accuracy, and overall performance.

Its immutable and decentralized ledger gadget ensures the integrity of statistics, facilitates real-time auditing, and allows computerized compliance through clever contracts. Blockchain also simplifies reporting techniques and enhances regulatory oversight with the aid of cozy and obvious information sharing and advanced traceability. Despite worrying conditions along with privacy worries and integration complexities, ongoing advancements in blockchain generation continue to deal with those issues, paving the way for its broader adoption in regulatory compliance and reporting.

## 6.8 FUTURE PROSPECTS OF BLOCKCHAIN IN BANKING

Blockchain technology has already made tremendous strides within the banking sector, supplying answers to diverse demanding situations consisting of security, performance, and transparency. However, its ability is some distance from exhaustion, and the future holds even extra promising possibilities for blockchain in banking. This segment explores the future directions of blockchain in banking, inclusive of rising traits, capability packages, and demanding situations for adoption.

### 6.8.1 Emerging trends

#### *6.8.1.1 Central Bank digital currencies (CBDCs)*

Central Banks International is exploring the issuance of Central Bank digital currencies (CBDCs), which can be virtual currencies issued and controlled by the central government. The blockchain era is natural in shape for CBDCs, providing relaxed and green structures for his or her issuance, distribution, and transaction processing. CBDCs have the potential to revolutionize the banking industry by offering faster, less expensive, and more inclusive payment systems (Bordo & Levin, 2017).

#### *6.8.1.2 Interbank settlement*

Blockchain-based structures like Ripple are gaining traction for interbank agreements, permitting faster and inexpensive cross-border payments. As more banks undertake blockchain for agreement functions, the traditional correspondent banking version can also become outdated, leading to greater performance and value savings (Peters & Panayi, 2016).

#### *6.8.1.3 Tokenization of assets*

Blockchain permits the tokenization of conventional assets, such as shares, bonds, and real property, turning them into digital tokens that can be traded on blockchain-based platforms. Tokenization can democratize access to funding opportunities, boost liquidity in asset markets, and streamline the issuance, buying, and selling of securities (Tapscott & Tapscott, 2016).

### 6.8.2 Potential applications

#### *6.8.2.1 Identity verification*

Blockchain can streamline identity verification methods in banking with the aid of secure and tamperproof virtual identities. Customers can control their right to access their identification data, reducing the risk of identification robbery and fraud. Blockchain-based identity answers should simplify customer onboarding, enhance protection, and enhance regulatory compliance (Zyskind, Nathan, & Pentland, 2015).

#### *6.8.2.2 Smart contracts for financial products*

Smart contracts enable the automation of monetary agreements, such as loans, derivatives, and insurance contracts. By encoding the phrases and conditions of those agreements into code, clever contracts can robotically execute and put in force contractual duties, decreasing the need for intermediaries, and minimizing the chance of disputes. Smart contracts may revolutionize the issuance and control of economic merchandise, making them extra efficient, transparent, and available (Christidis & Devetsikiotis, 2016).

#### *6.8.2.3 Regulatory compliance*

Blockchain can streamline regulatory compliance methods in banking by providing obvious and immutable data of transactions. Regulators can get entry to real-time records on banking activities, facilitating greater effective oversight, and decreasing the load of compliance reporting for banks. Blockchain-based compliance answers can enhance regulatory transparency, reduce compliance costs, and mitigate the hazard of regulatory breaches (Yermack, 2017).

## 6.9 CASE STUDIES AND REAL-WORLD APPLICATIONS OF BLOCKCHAIN IN BANKING

The blockchain era has won great traction inside the banking sector, with several real-world programs and case studies demonstrating its capacity to convert diverse aspects of banking operations. This segment examines some amazing case studies and real-world applications of blockchain in banking, illustrating how the technology is being utilized to decorate efficiency, security, and transparency.

### 6.9.1 J.P. Morgan's Interbank Information Network (IIN)

J.P. Morgan's Interbank Information Network (IIN) is a blockchain-based platform designed to streamline cross-border payments and compliance strategies for banks. By leveraging blockchain generation, IIN allows element banks to trade data securely and in real-time, lowering the time and charges associated with conventional correspondent banking networks. The IIN platform allows banks to affirm rate data straight away, lowering the chance of mistakes and delays in move-border transactions. Furthermore, IIN helps compliance with antimony laundering and recognizes your consumer pointers by presenting a cozy and immutable file of transaction statistics.

### 6.9.2 Santander's blockchain-based payment platform

Santander, one of the largest banks in Europe, has carried out blockchain generation to enhance its worldwide rate offerings. The financial group's blockchain-based total price platform, built in Ripple's technology, allows faster and additional fee-powerful move-border bills for its customers. By utilizing blockchain, Santander has decreased the time and prices associated with global transfers, offering customers an extra green and seamless fee enjoy. The use of blockchain technology has additionally more desirable transparency in fee processing, presenting clients with real-time visibility into the fame of their transactions.

### 6.9.3 HSBC's trade finance platform

HSBC, a worldwide banking massive, has carried out blockchain technology to digitize its alternative finance operations. The bank's blockchain-based trade finance platform, known as HSBC Connect, allows faster and more comfortable processing of alternative transactions, which include letters of credit score and invoices. By digitizing exchange finance approaches at the blockchain, HSBC has decreased paperwork and guide mistakes, main to extra efficiency and cost savings for each bank and its corporate customers. The transparency and immutability of blockchain technology also enhance belief and security in alternate transactions, lowering the chance of fraud and disputes.

### 6.9.4 Bank of America's blockchain patents

Bank of America, considered one of the largest banks in the United States, has been actively exploring blockchain generation and has filed several patents related to blockchain-based upgrades. These patents cover a huge range of programs, along with virtual forex structures, comfortable rate processing, and an information garage. One exceptional example is Bank of America's patent for a blockchain-based total system for wire transfers, which aims to improve the rate, safety, and transparency of wire transfer transactions. The patent describes a distributed ledger system that could facilitate the actual time agreement of cord transfers, decreasing the time and charges associated with traditional twine switch processes.

### 6.9.5 DBS Bank's digital exchange

DBS Bank, a main bank in Asia, has launched a virtual trade that leverages the blockchain era to permit tokenized asset buying and selling. The exchange permits institutions and buyers to exchange digital belongings which include cryptocurrencies, safety tokens, and virtual bonds in regulated and at-ease surroundings. By embracing the blockchain era, DBS Bank pursues to democratize get right of entry to digital property and offer buyers extra liquidity and transparency in asset trading. The virtual exchange additionally includes custody services and regulatory compliance capabilities, making sure of a secure and compliant buying and selling enjoy for traders.

## 6.10 CONCLUSIONS

Blockchain generation is unexpectedly reshaping the landscape of the banking company, providing transformative answers to age-old disturbing situations at the same time as introducing new possibilities for innovation. Throughout this exploration, we have witnessed how blockchain's decentralized ledger device, cryptographic safety, and clever agreement capabilities are revolutionizing numerous facets of banking operations. From enhancing safety and reducing fraud to streamlining strategies and allowing real-time transactions, blockchain has confirmed its ability to decorate performance, transparency, and consideration in banking. Case studies and actual international applications from main economic establishments like J.P. Morgan, Santander, HSBC, Bank of America, and DBS Bank underscore the tangible effect of blockchain on move-border payments, change finance, virtual asset buying and selling, and regulatory compliance.

Looking in advance, the opportunities of blockchain in banking are promising. Emerging trends such as CBDCs, tokenization of assets, and DeFi systems present new possibilities for blockchain innovation and adoption. However, disturbing conditions consisting of scalability, regulatory uncertainty, and privacy concerns remain sizeable boundaries that want to be addressed for blockchain to understand its whole capability in banking. As banks continue to spend money on blockchain studies and development, collaborate with agency companions, and interact with regulators, the vision of a decentralized, obvious, and inclusive monetary device will become increasingly plausible. By harnessing the energy of blockchain technology, banks can release new levels of efficiency, safety, and purchaser fees, ushering in a brand-new technology of banking innovation and prosperity.

## REFERENCES

Androulaki, E., Barger, A., Bortnikov, V., Cachin, C., Christidis, K., De Caro, A., ... & Yellick, J. (2018). Hyperledger Fabric: A Distributed Operating System for Permissioned Blockchains. *Proceedings of the Thirteenth EuroSys Conference*, 115. https://doi.org/10.1145/3190508.3190538

ASX. (2017). ASX to Develop Blockchain Solution for Equity PostTrade Processes. Retrieved from https://www.asx.com.au/content/dam/asx/about/media-releases/2017/ASX-Selects-DLT-to-Replace-CHESS-Media-Release-7December2017.pdf

Atzei, N., Bartoletti, M., & Cimoli, T. (2017). A Survey of Attacks on Ethereum Smart Contracts (SoK). *International Conference on Principles of Security and Trust*, 164186. https://link.springer.com/chapter/10.1007/978-3-662-54455-6_8

Azaria, A., Ekblaw, A., Vieira, T., & Lippman, A. (2016). MedRec: Using Blockchain for Medical Data Access and Permission Management. *2016 2nd International Conference on Open and Big Data (OBD)*, 2530. https://doi.org/10.1109/OBD.2016.11

Bank of America. (n.d.). Blockchain Patents. Retrieved from https://www.bankofamerica.com/company/patents/

Bordo, M. D., & Levin, A. T. (2017). Central Bank Digital Currency and the Future of Monetary Policy. *National Bureau of Economic Research Working Paper Series*, 23711. https://doi.org/10.3386/w23711

Buterin, V. (2013). Ethereum: A NextGeneration Smart Contract and Decentralized Application Platform. Retrieved from https://ethereum.org/en/whitepaper/

Buterin, V. (2014). A NextGeneration Smart Contract and Decentralized Application Platform. Retrieved from https://ethereum.org/en/whitepaper/

Castro, M., & Liskov, B. (1999). Practical Byzantine Fault Tolerance. *OSDI '99: Proceedings of the Third Symposium on Operating Systems Design and Implementation*, 173186.

Christidis, K., & Devetsikiotis, M. (2016). Blockchains and Smart Contracts for the Internet of Things. *The Plethora of Research in Internet of Things (IoT)*, 2292–2303.

Civic. (2018). Civic Secure Identity. Retrieved from https://www.civic.com/

Compound. (2020). Compound: The Money Market Protocol. Retrieved from https://compound.finance/

Croman, K., Decker, C., Eyal, I., Gencer, A. E., Juels, A., Kosba, A., ... & Wattenhofer, R. (2016). On Scaling Decentralized Blockchains. *International Conference on Financial Cryptography and Data Security*, 106125. https://link.springer.com/chapter/10.1007/978-3-662-53357-4_8

Crosby, M., Pattanayak, P., Verma, S., & Kalyanaraman, V. (2016). Blockchain Technology: Beyond Bitcoin. *Applied Innovation Review*, 2, 619.

Dai, J., & Vasarhelyi, M. A. (2017). Toward Blockchain-based Accounting and Assurance. *Journal of Information Systems*, 31(3), 521. https://publications.aaahq.org/jis/article/31/3/5/1105/Toward-Blockchain-Based-Accounting-and-Assurance

DBS Bank. (n.d.). DBS Digital Exchange. Retrieved from https://www.dbs.com/digibank/invest/dbsdigitalexchange.page

De Angelis, S., Aniello, L., Lombardi, F., & Margheri, A. (2018). PBFT vs ProofofAuthority: Applying the CAP Theorem to Permissioned Blockchain. *2018 1st International Conference on Cryptography, Security and Privacy (ICCSP)*, 16. https://ceur-ws.org/Vol-2058/paper-06.pdf?utm_source=chatgpt.com

Deloitte. (2017). Blockchain: Enigma. Paradox. Opportunity. Retrieved from https://assets.ctfassets.net/sdlntm3tthp6/resource-asset-r120/d27ab8f9d99d62525e79b1b6ada2b822/9c9e68f6-5c38-4ff6-9d9e-e84008417b8a.pdf

DTCC. (2017). Embracing Disruption: Tapping the Potential of Distributed Ledgers to Improve the PostTrade Landscape. Retrieved from https://www.dtcc.com/

European Union Blockchain Observatory and Forum. (2019). Legal and Regulatory Framework.

Everledger. (2019). Everledger: Provenance Tracking for HighValue Goods. Retrieved from https://www.everledger.io/

FDA. (2019). FDA's New Era of Smarter Food Safety: Using Blockchain for Drug Traceability. Retrieved from https://www.fda.gov/

Felin, T., & Lakhani, K. (2018). What Problems Will You Solve with Blockchain? *MIT Sloan Management Review*, 60(1), 1820.

Gupta, M. (2017). *Blockchain for Dummies*. John Wiley & Sons.

Harvey, C. R., Ramachandran, A., & Santoro, J. (2021). DeFi and the Future of Finance. Retrieved from https://papers.ssrn.com/sol3/papers.cfm?abstract_id=3711777

Kamath, R. (2018). Food Traceability on Blockchain: Walmart's Pork and Mango Pilots with IBM. *The Journal of the British Blockchain Association*, 1(1), 4753. https://jbba.scholasticahq.com/article/3712-food-traceability-on-blockchain-walmart-s-pork-and-mango-pilots-with-ibmPDF: https://jbba.scholasticahq.com/article/3712-food-traceability-on-blockchain-walmart-s-pork-and-mango-pilots-with-ibm/attachment/20459.pdf

King, S., & Nadal, S. (2012). PPCoin: PeertoPeer CryptoCurrency with ProofofStake. Retrieved from https://decred.org/research/king2012.pdf

Kosba, A., Miller, A., Shi, E., Wen, Z., & Papamanthou, C. (2016). Hawk: The Blockchain Model of Cryptography and PrivacyPreserving Smart Contracts. *2016 IEEE Symposium on Security and Privacy (SP)*, 839858. https://doi.org/10.1109/SP.2016.55

Larimer, D. (2014). Delegated ProofofStake (DPoS). BitShares. Retrieved from https://bitshares.org/delegated-proof-of-stake-consensus/

Ledger. (2019). MediLedger Network. Retrieved from https://www.mediledger.com/

Loyyal. (2019). Loyyal: Blockchain for Loyalty Programs. Retrieved from https://www.loyyal.com/

Marr, B. (2018). How Blockchain Will Transform the Insurance Industry. Forbes. Retrieved from https://www.forbes.com/sites/bernardmarr/2020/08/05/fascinating-examples-of-how-blockchain-is-used-in-insurance-banking-and-travel/?utm_source=chatgpt.com

MediBloc. (2018). MediBloc: Decentralized Healthcare Information Ecosystem. Retrieved from https://medibloc.org/en/

MediLedger. (2019). MediLedger Network. Retrieved from https://www.mediledger.com/

Mettler, M. (2016). *Blockchain Technology in Healthcare: The Revolution Starts Here*. IEEE.

Morabito, V. (2017). *Business Innovation Through Blockchain: The B3 Perspective*. Springer.

Mougayar, W. (2016). *The Business Blockchain: Promise, Practice, and Application of the Next Internet Technology*. Wiley.

Nakamoto, S. (2008). Bitcoin: A PeertoPeer Electronic Cash System. Retrieved from https://bitcoin.org/bitcoin.pdf

Peters, G. W., & Panayi, E. (2016). Understanding Modern Banking Ledgers through https://doi.org/10.1007/978-3-319-42448-4_13

Pilkington, M. (2016). Blockchain Technology: Principles and Applications. In F. X. Olleros & M. Zhegu (Eds.), *Research Handbook on Digital Transformations* (pp. 225–253). Cheltenham: Edward Elgar Publishing https://doi.org/10.4337/9781784717766.00019.

Provenance. (2015). Blockchain: The Solution for Transparency in Product Supply Chains. Retrieved from https://www.provenance.org/

Ripple. (2018). Ripple: One Frictionless Experience to Send Money Globally. Retrieved from https://ripple.com/

Ripple. (2018). Santander Launches First Blockchain-Based International Money Transfer Service Across Four Countries. Retrieved from https://www.santander.com/content/dam/santander-com/en/documentos/historico-notas-de-prensa/2018/04/NP-2018-04-12-Santander%20launches%20the%20first%20blockchain-based%20international%20money%20transfer%20service%20across%20-en.pdf

Saberi, S., Kouhizadeh, M., Sarkis, J., & Shen, L. (2019). Blockchain Technology and Its Relationships to Sustainable Supply Chain Management. *International Journal of Production Research*, 57(7), 21172135. https://doi.org/10.1080/00207543.2018.1533261

Shah, R. (2017). Blockchain Voting: A New Paradigm for Modern Democracy. *Journal of Political Science*, 5(2), 7894.

Shen, M., Guo, J., & Yang, Y. (2019). MedChain: Efficient Healthcare Data Sharing via Blockchain. *Applied Sciences*, 9(6), 1206. https://doi.org/10.3390/app9061206

Swan, M. (2015). *Blockchain: Blueprint for a New Economy*. O'Reilly Media.

Szabo, N. (1997). Formalizing and Securing Relationships on Public Networks. *First Monday*, 2(9). https://doi.org/10.5210/fm.v2i9.548

Tapscott, D., & Tapscott, A. (2016). Blockchain Revolution: How the Technology Behind Bitcoin Is Changing Money, Business, and the World. *Penguin. of the ACM*, 59(11), 1517. https://doi.org/10.1145/2994581

Yermack, D. (2017). Corporate Governance and Blockchains. *Review of Finance*, 21(1), 731. https://doi.org/10.1093/rof/rfw074

Zcash. (2016). The Zcash Protocol Specification. Retrieved from Working paper on SSRN: https://papers.ssrn.com/sol3/papers.cfm?abstract_id=2700475

Zheng, Z., Xie, S., Dai, H., Chen, X., & Wang, H. (2017). An Overview of Blockchain Technology: Architecture, Consensus, and Future Trends. *2017 IEEE International Congress on Big Data (BigData Congress)*, 557564. https://doi.org/10.1109/BigDataCongress.2017.85

Zohar, A. (2015). Bitcoin: Under the Hood. *Communications of the ACM*, 58(9), 104113. https://doi.org/10.1145/2701411

Zyskind, G., Nathan, O., & Pentland, A. (2015). Decentralizing Privacy: Using Blockchain to Protect Personal Data. *2015 IEEE Security and Privacy Workshops*, 180184. https://doi.org/10.1109/SPW.2015.27.

Chapter 7

# Blockchain disruptive role for empowering cyber security

*Medini Gupta, Sarvesh Tanwar, and Ravindra Kumar Singh Rajput*

| Abbreviations | Definition |
|---|---|
| PKI | Public Key Infrastructure |
| IoT | Internet of Things |
| AI | Artificial Intelligence |
| DDoS | Distributed Denial of Service |
| IDS | Intrusion Detection System |
| CYBEX | Cybersecurity Information Exchange |
| PoW | Proof of Work |
| PoS | Proof of Stake |
| DPoS | Delegated Proof of Stake |
| PoA | Proof of Authority |

## 7.1 INTRODUCTION

Technology has brought vast development at an unparalleled scale. Innovative solutions have led to paradigm shift in our daily lifestyle that was not heralded few decades back.

Although blockchain is in its evolving stage, it has shown enormous potential in recent times. Blockchain is a radical form of electronic ledger that logs information in a highly secure and immutable environment [1]. Decentralization and cryptographic techniques are implemented to achieve high security of records. These key parameters are responsible for establishing a counterfeit-free blockchain network. Bitcoin cryptocurrency is the backbone that led to the birth of blockchain. Since then, this technology has been adopted to revolutionize various other sectors that includes identity theft management, non-fungible transfers, asset tokenization, intellectual property data and many more. We will discuss about its use cases later in this chapter.

DOI: 10.1201/9781003504559-7

Blockchain operates by organizing the data into different blocks that are associated with each other cryptographically. Crypto contract, more commonly termed as smart contract are computerized programs that automatically govern the transmission of digital resources among the entities under specific criteria [2]. Conventional contracts are imposed by law whereas smart contracts are imposed by the code. Figure 7.1 represents the workflow of blockchain. The decentralized ledger gets updated in real time, with new additional blocks in every 10 minutes on an average. Consensus algorithms are utilized to decode complicated numerical puzzles on the basis of which new blocks are included. Before adding new transaction on the chained network, all the transactions are approved by miners on the platform that verify the validity of the transaction by solving tricky numerical puzzles. Scalability is still an important challenge for organizations as performance of the ledger degrades when number of transaction rises [3]. Much more computational energy is required to verify the records. High latency and low speed make it inappropriate for managing real-time transactions. To meet the needs of the organizations, there should be proper establishment of legal framework. Bugs on the smart contract can be misused leading to unauthenticated access, privacy and economical damage. Decentralization feature of blockchain states that no single authority or body controls the ledger, rather it depends on computers all over the world to retain its integrity and security [4].

Cybersecurity is the procedure that deals with safeguarding network servers, data and electronic devices such as computers, tablets, smart phones from malicious threats by detecting any flaws in the network. Globally cyber threats are evolving at an accelerating rate. Along with monetary cost, organizations also experience non-monetary loss, like reputational sabotage. Employing appropriate cybersecurity measures is very much crucial for defending computer networks from cyber threats. Different hardware

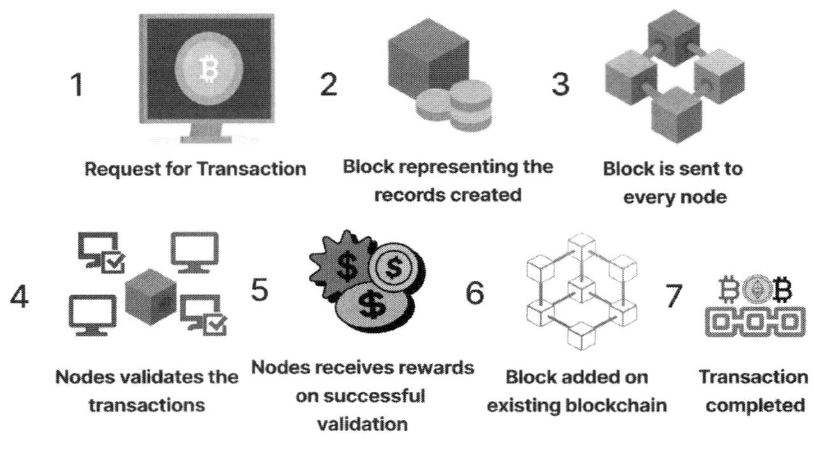


*Figure 7.1* Working of blockchain.

and software are designed to protect web assets from falling prey to cyber vulnerabilities. Suitable security policies are deployed based on three interconnected and foundational concepts in cyber world: Confidentiality, Integrity and Availability [5]. Confidentiality initiatives are built to prevent leakage of sensitive details from illegitimate entities. In integrity, measures are undertaken to ensure that data doesn't get altered during its lifecycle. Availability deals with accessibility of assets to authorized individuals by keeping technical resources and hardware up to date. Confidentiality is parallel to privacy, integrity with trustworthiness of information whereas availability is equivalent to consistency [6].

Cyber defense cannot be considered as 100% full proof secure, as the network which is regarded as safe in present might not be so information stored are pertaining to ingenuine nature of criminals for seeking different methodologies to perform cyberattacks. Blockchain guides a new route towards higher security area, that is rarely traveled as hospital for malicious attackers. A distributed public key infrastructure (PKI) that mitigates lots of risk connected to information that is stored on central systems [7]. With the exception of vulnerability exposure, it is challenging for cyber attackers to compromise with confidential data on the distributed ledger as all the transactions are unrecorded among the nodes on the platform. Consensus algorithm on blockchain platform looks for illegal activities and faults without involving central authority.

### 7.1.1 Smart contract to mitigate cyberattacks

Smart contract is one of the crucial components of blockchain that allows the development of auto-execution of agreements between the buyer and seller respectively. These are computerized bonds that run quickly and automatically when prespecified criteria are accomplished. Smart contracts are kept on the blockchain that delivers auditable, immutable and tamper-resistant records [8]. These contracts streamline the digital transactions process by mitigating the entry of intermediaries and particularly reducing data manipulation. Thereby increasing the reliability of digital assets from malicious entities. Traditional contracts usually suffer from man-made errors whereas the accuracy of smart contracts is much higher, owing to the fact that smart contracts execute only when pre-assumed terms and conditions get fulfilled. Parties involved in the smart contract don't need to keep faith on each other to undertake contractually suitable steps because these contracts are self-governing and eradicate the presence of third parties like middlemen and brokers. Eliminating intermediaries proves economically feasible for organizations as they don't need to pay them for facilitating transactions. Records stored on the smart contract are encrypted as they run on a peer-to-peer network and thus are less susceptible to fraud [9]. Each participant of a smart contract has live access to records, which boosts the transparency and immutability of the transactions.

### 7.1.2 Motivation

With increased reliance on technology, it has become very much significant to strengthen security regulations in order to safeguard web transactions and data. Malicious threats can be conducted using numerous malwares including rootkits, viruses and trojans. Blockchain maintains real-time records of transactions being performed in a sequence of blocks. Immutability is one of the key features of this chained network that regulates transparency of information. The usage of Internet of Things (IoT) and Artificial Intelligence (AI) smart appliances are still at its peak. These appliances also generate a huge volume of data [10]. Protection of data and appliances from illegitimate hackers is a matter of serious concern. Blockchain is capable of enhancing the security of smart appliances by carrying out end-to-end encryption of every device. Distributed Denial of Service (DDoS) is the most prominent type of cyber threat where attackers target to produce flood of cyberspace traffic and hinder the service flow. Blockchain implies public key infrastructure and decentralized measures to mitigate cyber threats.

### 7.1.3 Research questions

**RQ1. What are the key principles of blockchain through which cybersecurity can be integrated successfully?** By utilizing distributed and decentralized features of the ledger, organizations can boost transparency, expand security and safeguard confidential information from illegitimate control and manipulation [11].

**RQ2. What are the different use cases of blockchain which can be revolutionized while maintaining a threat-free cyber landscape?**

Blockchain enables record tracking of goods and services at each and every step of supply chain. Transactions performed on the decentralized ledger are everlasting and can't be updated afterwards, thus delivering trustworthy course of activities.

**RQ3. Is there any cyber threat associated with blockchain that is hindering its smooth workflow?** In a blockchain sybil attack, fraudsters create multiple duplicate identities to gain unauthorized access and manipulate the records. Because of inappropriate blockchain implementation, security threats have risen that consistently affect the development of project with their stakeholders.

**RQ4. How can cyber vulnerabilities be eradicated to ensure security on the decentralized ledger?** Consensus algorithms can be treated as a mechanism through which peers on the distributed ledger come to a consensus about the actual status of the blockchain [12]. There are multiple consensus algorithms that work toward attaining unified agreement.

## 7.2 LITERATURE REVIEW

Alex. R. Mathew [1] explored the impact of blockchain on cybersecurity sector by performing a qualitative review of secondary data. Author then analyzed the applications of blockchain in securing IoT. Even though existing technical solutions retrain a degree of defensible measures for computer resources, they are susceptible for one-time failure. This can be due to the fact that these many of security mechanisms are executed to run autonomously while defending computer infrastructure. Blockchain, being distributed ledger, records its details on multiple nodes instead of securing them on a single server. Major concern for IoT security is towards unauthorized control to IoT smart devices. Blockchain can hinder these issues in a more appropriate way by implementing a consensus protocol that would verify the new end user. A trustworthy and authentic user recognition, and data transmission system can be built with blockchain.

Mounir Frikha et al. [2] delivered a comprehensive study on cybersecurity with blockchain-based IoT devices. Authors examined the potential benefits of IoT with immutable ledger such as integrity of data, decentralized and transparent solution. However, they have also mentioned about the rising challenges pertaining to blockchain security, IoT device security and convergence of blockchain security with IoT. With a finite number of resources, low level of computational capacity and storage ability, IoT smart devices are mostly prone to cyber vulnerabilities such as tampered data, malicious attacks, security breaches and illegal access control. Blockchain implies consensus protocols to perform verification of transactions as well as users. Consensus protocols assure that all the participants present on the platform agree with the validation procedure. Cryptographic methods make unauthorized information unreadable for the users.

Atharva Deshmukh et al. [3] in their survey paper discussed about the security mechanism of blockchain-based cybersecurity solution. It is examined that encryption, decentralization, authorized access and immutability characters of blockchain would assist in achieving a secured environment for various systems. In IoT, mainly two categories of authentication take place. Firstly, authenticating users to the smart devices and then verifying those devices with the network system. Moreover distributed, profitable and decentralized crypto tokens are powerful and secure procedures for generating revenue. The demand for electricity in vast amount to power up all the computer infrastructure, skillful individuals suitable to work with this ledger, other technologies that are competitors of blockchain posse some challenges for the complete acceptance of blockchain.

Samreen Mahmood et al. [4] in their study mentioned about the challenges of cybersecurity with blockchain technology, where the authors talked about high energy consumption is required by blockchain to deliver real-time information with transparency. Blockchain system needs considerable amount of investment in initial stage with threat of external breaches including sybil's attack, 51% attacks and double spending vulnerability. Sybil attack occur when proprietor create various duplicate identities to create computer resources, in order to underestimate authorities. In spite of the fact that blockchain adoption has risen up because of its unique characteristics, study opens up about the security-related issues in ledger. Smart contract loophole vulnerability, wallet security threat and system error vulnerability are the most faced attacks on the decentralized blockchain.

## 7.3 BLOCKCHAIN ADVANCEMENT TO ENHANCE CYBERSECURITY

Blockchain has earned much significance in past years because of its inherited features that have played crucial role in revolutionizing various sectors, and cybersecurity is also one among them [13]. By utilizing distributed and decentralized features of the ledger, organizations can boost transparency, expand security and safeguard confidential information from illegitimate control and manipulation. This section will cover applications of blockchain for cyber security.

### 7.3.1 Reliable audit trails

Retaining a trustworthy audit trail is important for conserving the integrity of cybersecurity procedures [14]. Immutable audit trails of blockchain provides a perfect solution for developing tamper-resistant audit records. Transactions performed on the decentralized ledger are everlasting and can't be updated afterwards, thus delivering a trustworthy course of activities. By implementing blockchain for audit trails, about 30% of cost reduction can be achieved by mitigating the requirement of traditional conciliation [15].

### 7.3.2 Comprehensive threat intelligence

Threat Intelligence Sharing is a broader way to deal with malicious vulnerabilities of cyberspace. Cyber threats information is directly shared among the entities. Information about the threat is gathered, examined and then shared by the respective organizations that are involved in the threat intelligence sharing group [16]. Threat intelligence is coupled with a number of benefits to safeguard an organization's network infrastructure but also has a reasonable number of challenges in terms of maintaining privacy and ensuring that threat intelligence does not reveal personal or confidential information. Countering cyber vulnerabilities need broad and real-time threat intelligence [17]. With blockchain, organizations can smoothly share threat details with their trustworthy parties while maintaining obscurity and confidentiality of the associates involved (Figure 7.2).

*Table 7.1* Review on blockchain solutions

| *Authors* | *Study* | *Existing challenges* | *Proposed architecture* |
|---|---|---|---|
| Tarek Saadawi et al. [5] | Blockchain-based Intrusion Detection System (IDS) | Distributed attacks are not detected on time despite increasing the accuracy of intrusion detection system.<br>Security of the data being shared is not assured. | Proposed solution identifies and also prevents duplicate feature injection, malicious deletion. Maintains the integrity and consistency of attack characteristics distributed on a cooperative IDS. Provides IDS nodes with scalable cyberattack features exchange and promises diversified involvement of IDS nodes. It is scalable with low-latency IDS nodes. |
| Shamik Sengupta et al. [6] | Ethereum Blockchain-based BlocCyNfo-Share model | Security vulnerabilities in centralized Cybersecurity Information Exchange that undergoes complete downtime due to any causality. Furthermore, while exchanging sensitive details, it has absence of proper mechanism for query organization. Non-repudiation i.e., record tracking about the data being shared within the organization does not exist. | Blockchain-based privacy preserving cybersecurity information sharing using proxy re-encryption and retribute based encryption model, here establishments can share their data regarding cyber threat intelligence while maintaining privacy. The protocol does not have time frame and delivers two levels of fine stage access. In the first stage user can select the respective organization. Which will receive the encrypted data. Organization's identity is not reversed when they take exist from the organization. Re-encryption key, that is used to re-encrypt network access server gets revoked, making the system more flexible. Tamper-proof data is achieved through immutability of blockchain, through which owner cannot compromise with integrity of data that Is already stored safely on ledger with time stamp. |
| Jamilu Sabi'u et al. [7] | Ethereum blockchain for smart city-based organizations | Risk pertaining to transactional privacy infringement in wide private network triggered by hackers by employing their exploitation capabilities. | Blockchain with digital certificates, public key infrastructure through RSA to achieve security and privacy for wide private network. Encryption speed in this model has increased by 17 times whereas decryption has accelerated by 41 times. The information transmitted remains confidentiality and integrity in node-to-node communication within wide private system. |

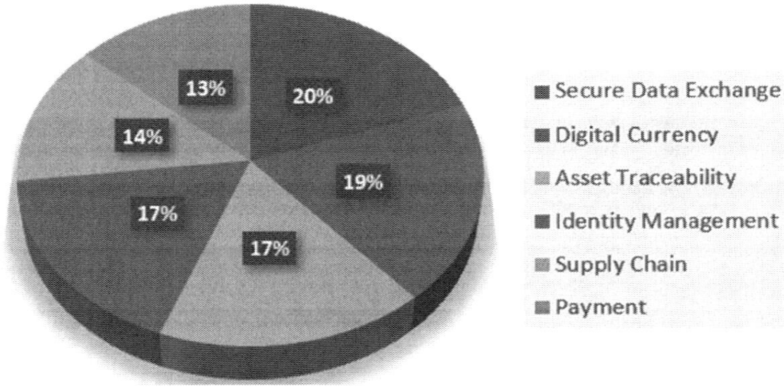


*Figure 7.2* Popular use cases of blockchain in 2022 [20].

### 7.3.3 Guaranteed identity verification

Conventional solutions usually depend on database centralization that are commonly prone to security breaches and hacking. Identity verification is one of the major challenges of cybersecurity [18]. Blockchain works on the principle of decentralization for identity verification. This approach of blockchain allows entities to have complete access control and their sensitive details are kept securely on a distributed ledger.

### 7.3.4 Securing supply chain

Blockchain enables record tracking of goods and services at each and every step of supply chain [19]. Visibility of supply chain procedure from farm to fork is achieved in transparent and decentralized manner. Counterfeiting and tampering at any stage of the journey can be quickly identified. Time-taking confirmations can be replaced by encapsulating smart contracts, that spontaneously updates the activities when prespecified terms are fulfilled.

## 7.4 POTENTIAL CYBER THREATS IN BLOCKCHAIN

Regardless of high opportunities of implementing blockchain in diverse fields, it is also coupled with its unfavorable impacts. Considering that blockchain foundation is laid on cryptographic infrastructure, it is generally mistaken as fully secure by technological design [20]. But these inbuilt characteristics are not enough to combat existing cyber threats [21]. Blockchain features of openness and decentralization are responsible for increasing the complication of operations and brought down the chances of gaining entire access control, leading towards in-depth examination to safeguard against digital threats. As every new technology brings its own techniques for cyber exploitation, blockchain is also among them. Depending on the motive, blockchains are developed in which end users have multiple access [22].

Public blockchain allows anyone connected with network to participate and this type of blockchain is permissionless that promotes transparency, decentralization and untrustworthy transactions. Private blockchain are permissioned and restrictive in nature that strengthen the privacy and efficiency of operations on the network by enabling only verified members to edit, update or delete transactions [23]. Consortium blockchains are a convergence of private and public blockchain that are partially decentralized. Because of inappropriate blockchain implementation, security threats have risen, consistently affecting the development of projects with its stakeholders [24]. Below we have discussed about the blockchain security attacks that have caused considerable amount of havoc on blockchain system.

### 7.4.1 Double spending attack

In double spending attack, there is a threat of using cryptocurrency twice or more for the single transaction. Details of the transaction can be altered by fraudster within the blockchain if the particular terms and conditions are fulfilled. Those conditions allow updated blocks to breach the security of blockchain, thus enabling that fraudster to retrieve crypto coins that were already spent [25]. When the cryptocurrency is not stored securely on the wallet, then it can get stolen from the ledger. While creating a block on the ledger, an encrypted value that comprises of transaction details, time stamp and data about previous block is received. This data is being encrypted by applying security mechanism such as SHA-256 used in Bitcoin transaction [26]. Miners then authenticate the block details and closes that particular block. Again, a new block is created with hash of previous block, time stamp and transaction data. The miner whose machine successfully authenticates crypto hash gets bitcoin as a reward. For performing double spend attack, fraudster has to mine a hidden block that will surpass the development of original blockchain. Then the hidden block that is mined is introduced to the original blockchain network [27]. If the fraudster successfully performs this then the new block is added to the chained network. Figure 7.3 represents illegal share of crypto transactions between 2017 and 2022. Through this attack, an illicit individual is able to get hold of cryptocurrencies that were already spent.

### 7.4.2 51% Cyber attack

This attack is undertaken by various miners that hold more than up to 50% of network hash rate by clearing the complex numerical puzzle. Those miners then bring up updated blockchain to, that gets acknowledged by the

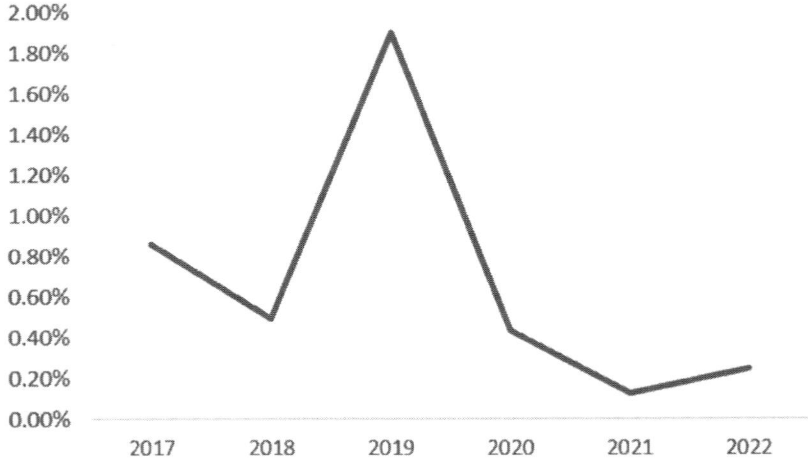


*Figure 7.3* Illegal crypto transactions between 2017 and 2022 [24].

network. Possessing control of 51% of the nodes on the ledger allows entities to perform alterations on the network [28]. Illicit entities gain power to terminate transactions between the end users, hamper the smooth workflow of transactions being undertaken and even reverse records that were already stored. This attack also leads to a double spending threat. Fraudster being in the majority can divert the transactions by prohibiting rest of the miners from finishing their blocks [29]. Majority consensus about every transaction is performed for validation purpose. The blocks where details are kept are locked or sealed. Transactions that are sealed before the attack occurred are impossible to breach because of immutability of blockchain. When the participants on the platform are high then it is more challenging to perform a 51% attack on the cryptocurrency [30].

### 7.4.3 Sybil attack

In Sybil attack, individuals have multiple duplicate identities for malicious reasons on the network. Nodes underestimate the power of blockchain by maintaining various fake IDs at the same time interval in order to achieve a majority rate to execute inappropriate actions [31]. A computer can develop multiple identities like IP addresses and user accounts. Systems and end users are tricked in assuming these duplicate ids as original ones. When fake actors are able to outsmart blockchain then they can also disallow original nodes to perform transactions. This often causes a 51% attack where integrity of the network is put on stake. Validation of identities can be fruitful for preventing sybil attack by disclosing the true identity of fake actors. Direct verification takes place when central governance performs verification of remote identities [32]. In indirect validation, local bodies depend on previous identities that were accepted, through this rest of the nodes certify the validation of remote identities.

### 7.4.4 DDoS attack

Blockchain being distributed in nature enables the smooth flow of transaction even when some of the nodes are not connected to network. Any node can be offline due to Distributed Denial of Service (DDoS) attack or because of another issue without hindering the entire network. The immutable ledger is not resistant to DDoS attack. The robustness of the chained network entirely depends on its hash rate, total number of nodes connected and diversity. Attackers can compromise the security of the system by flooding the network with spam and fake transactions. Blockchain have its pre-specified capacity as additional blocks are built at regular pace [33]. The records which are not included in the existing block are transmitted to the Mempool, where they will be included in an upcoming block. When fraudster transmits transactions in bulk then the entire block will be occupied with fake and spam records leading valid transactions to remain in the Mempool for longer duration. The downtime price of IT resource is much high, along with the adverse effect it will have on revenue and brand reputation.

## 7.5 CASE STUDIES

Instant money is transmitted through digital bank accounts. Nowadays online orders are also delivered very next day. But these conveniences in our routine life have come with a huge cost. The web contains a massive number of accounts that are secured by weak passwords that includes passport details, bank accounts, social accounts and health details. Any gap in security will expose our confidential details in one go. Establishing a trustworthy cyber protocol is the need of the hour [34]. Different industries are on the board have put their hopes on emerging technologies that ensure enhanced web security, with cybersecurity and blockchain showing a trusted path. Below, we have covered emerging case studies of blockchain in different industries.

### 7.5.1 Philips Healthcare

Philips Healthcare is a subsidiary of Philips Research Organization, has coupled AI with blockchain to develop health infrastructure. It has built partnerships with hospitals around the world [35]. The organizations utilize AI to identify and examine different angles of healthcare such as surgical operations, health records and administration. Blockchain is then implemented for securing enormous volume of data being gathered.

### 7.5.2 Health linkages

Health Linkages based on California implements an immutable ledger to deliver tamper-proof data administration, thus increasing reliable audit logs and conformity in the health domain. This blockchain solution only allows credentialed participants to share records of their patients, and it also retains health events of their patients in chronological order. This approach gives doctors a clear view of patient's health status [36].

### 7.5.3 Lockheed Martin

Lockheed Martin is the initial defense organization that executed blockchain. Along with Guardtime Federal, they executed blockchain-based cybersecurity protocol in software growth, risk monitoring in the supply chain and engineering systems [37]. Consortium blockchain is being used to make its cyber security capabilities more vigorous.

### 7.5.4 Xage Security

To cater to the requirement for enhancing IoT ecosystem, Xage, located in California, has developed blockchain platform that is associated with highly advanced security protocols. Only the members of the blockchain can perform legitimate transactions, maintaining a zero-trust remote access strategy. Fraudsters don't have the opportunity to take advantage of inadequately mentioned permissions [38]. In this way organizations can safeguard the records that are transmitted among smart devices.

## 7.6 CONSENSUS MECHANISMS TO TACKLE CYBER THREATS

Blockchain consensus mechanism is used to gain security and reliability in a decentralized architecture. It plays a pivotal role in securing the transactions by performing encryption and executing smart contracts. It can be treated as a mechanism through which peers on the distributed ledger come to a consensus about the actual status of the blockchain. Consensus algorithms are capable of replacing slower manual validation procedure. These algorithms look for the size of the block, timestamp of block, first transaction recorded, header hash of block and data structure being used and on the basis of information gathered, finalize the state of the records [8]. Trust is established among the nodes on the platform by making sure that the ledger is free from security threats. There are multiple consensus algorithms that work on attaining unified agreement. The real-time status of the blockchain should be up to date is achieved through these algorithms.

### 7.6.1 Proof of Work

Proof of Work (PoW) is the popular consensus algorithm implemented in various peer-to-peer networks to verify transactions and adding newly created blocks on the ledger. Initially, PoW was executed on bitcoin cryptocurrency in order to safeguard the network from double spending and 51% attack. For verifying and adding new blocks, miners have to successfully resolve complicated numerical puzzles termed as hashes. PoW hash function is designed in such a way that miners have to put a lot of computational ability, before creating additional blocks. Miners compete to resolve the hash and, the one who resolves the hash function first among them is awarded with cryptocurrency [9]. Figure 7.4 shows the working of PoW. A massive amount of energy is absorbed to solve numerical puzzle. The security of this mechanism lies on the complexity of the hash, as it would be nearly impossible for fraud actor to breach the network. For the fraud actor to execute their illegal intentions, they must have control over 51% of hash rate.

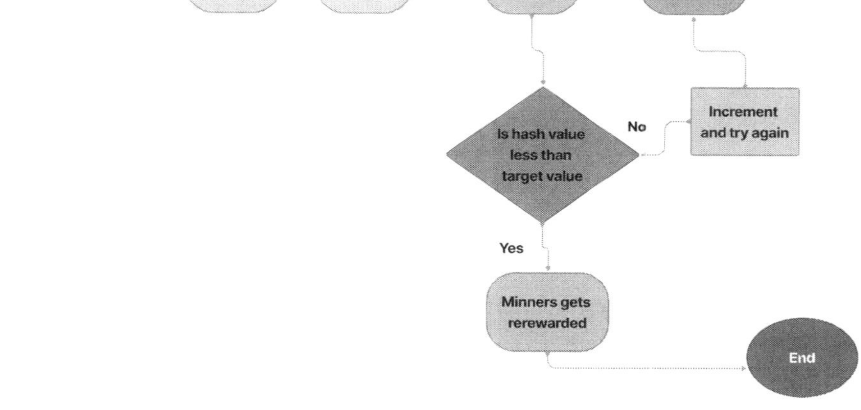


*Figure 7.4* PoW workflow.

### 7.6.2 Proof of Stake

Proof of Stake (PoS) came into existence to overcome the challenge of high energy consumption of PoW algorithm. This mechanism is also used for validating and adding new blocks. In PoW, miners are required to solve complicated puzzle whereas in PoS, coin possession is the criteria for choosing validators [10]. The one who holds significant quantity of cryptocurrencies called stakes, are selected for performing validation. Validators are motivated to remain trustworthy as they are likely to lose their stake, if they try to breach network security or verify inappropriate transaction. As PoS absorbs low energy, it is less expensive and more eco-friendly. PoS encourages decentralization as miners with highest number of crypto possessions have control and it is impractical for single or multiple individuals to earn control of ledger. Because as a means to do so, they have to control a large quantity of crypto coins. PoW is more prone to cyber vulnerabilities with miners having high hash rate can control the chained network. The major drawback with PoS is that the validator who have less stake would hardly get the opportunity to validate and build new blocks and on the other side, the validators with large stake will continue to earn more.

### 7.6.3 Delegated Proof of Stake

Delegated Proof of Stake is the subcategory of PoS, that involves a limited number of miners, termed as witnesses or delegates. These delegates perform transaction verification and build new blocks. Users or token holders select delegates to verify upcoming block. To elect a delegate, tokens are transmitted to staking pool and those tokens are interlinked with that specific delegate. The elected witness gets rewarded with transaction amount from verified block and thereafter, that incentive is being distributed among the token holders who pooled with the elected delegate [11]. The higher the amount you stake, the higher you will earn. Between 20 and 100 delegates are chosen so that witness of one block doesn't repeat in the next block. With limited number of witnesses, Delegated Proof of Stake quickly reaches consensus.

### 7.6.4 Proof of Authority

Proof of Authority lays the foundation on trustworthy miners. In this mechanism, a group of miners is given the responsibility of record verification and to build new blocks on the ledger. Miners are elected depending on their experience and reputation. Miners are more focused because their image is at stake. PoS needs a large number of stakes to enroll, which can cause centralization [12]. With PoW, high energy consumption on mathematical puzzles for validating records, makes this algorithm expensive and time-consuming. On the other hand, PoW involves only limited and authentic miners that makes the procedure fast and smooth. Proof of Authority is susceptible to cyber threats if any validator gets compromised.

## 7.7 CONCLUSION AND FUTURE SCOPE

Blockchain has brought a complete turnover in reshaping the future of cyber security, encouraging organizations with additional transparency, trust and asset regulation [2]. Cyber breaches have emerged as the leading threat on the digital landscape, resulting economic and reputational loss. Conventional security protocols are not accurate in mitigating threats, leading technical experts to look for alternative and innovate solutions [1]. A timestamp ledger have promised to deliver threat-proof alternative of centralized storage by introducing decentralization of records and pairing it with encryption algorithms. By embracing cryptographic hashing techniques, blockchain achieves immutability of records stored on the ledger securely. We have discussed about different use cases of blockchain in supply chain, identity management traceability, trustworthy auditable trails. By utilizing the capabilities of blockchain, organizations can eradicate attacks and can protect their online assets. Smart contracts automatically run when preestablished terms are fulfilled, that also brings down human errors and speeds up the transaction process.

Blockchain architecture, like PoW consumes a significant extent of energy while mining the transactions. The prominent challenge for blockchain is its scalability, the block sizes also increase along with the total number of transactions on the ledger, leading to high latency and low speed [4]. Consensus algorithm demands for more energy consumption where entities work on complicated numerical puzzles for validation process. Blockchain is not a panacea; it is also susceptible for cyber threats like 51% attacks, DDoS attacks, sybil attack. Intersection of blockchain and cybersecurity can protect confidential information and hamper illegitimate access by making use of persistent cryptographic algorithms. As blockchain has gained widespread attention of organizations in most of the sectors, it is crucial to outsmart malicious actors by applying appropriate secure mechanisms. There is immense scope for development in blockchain and cybersecurity sector.

## REFERENCES

[1] Mathew, Alex R. "Cyber security through blockchain technology." *International Journal of Engineering and Advanced Technology* 9, no. 1 (2019): 3821–3824.

[2] Alajlan, Razan, Norah Alhumam, and Mounir Frikha. "Cybersecurity for blockchain-based IoT systems: A review." *Applied Sciences* 13, no. 13 (2023): 7432.

[3] Deshmukh, Atharva, N. Sreenath, Amit Kumar Tyagi, and Uppalaguptapu Venkata Eswara Abhichandan. "Blockchain enabled cyber security: A comprehensive survey." In *2022 International Conference on Computer Communication and Informatics (ICCCI)*, pp. 1–6. IEEE, 2022.

[4] Mahmood, Samreen, Mehmood Chadhar, and Selena Firmin. "Cybersecurity challenges in blockchain technology: A scoping review." *Human Behavior and merging Technologies* 2022 (2022): 1–11.

[5] Ajayi, Oluwaseyi, and Tarek Saadawi. "Blockchain-Based Architecture for Secured Cyber-Attack Features Exchange." In *2020 7th IEEE International Conference on Cyber Security and Cloud Computing (CSCloud)/2020 6th IEEE International Conference on Edge Computing and Scalable Cloud (EdgeCom)*, pp. 100–107. IEEE, 2020.

[6] Badsha, Shahriar, Iman Vakilinia, and Shamik Sengupta. "Blocynfo-share: Blockchain based cybersecurity information sharing with fine grained access control." *2020 10th Annual Computing and Communication Workshop and Conference (CCWC)*. IEEE, 2020.

[7] Yakubu, Bello Musa, Jamilu Sabi'u, and Pattarasinee Bhattarakosol. "Blockchain-based privacy and security model for transactional data in large private networks." *Scientific Reports* 13, no. 1 (2023): 17108.

[8] Kakkar, Latika, Deepali Gupta, Sapna Saxena, and Sarvesh Tanwar. "An analysis of integration of internet of things and cloud computing." *Journal of Computational and Theoretical Nanoscience* 16, no. 10 (2019): 4345–4349.

[9] Gupta, Medini, Sarvesh Tanwar, Niranjan Lal, and Ritvik Pawar. "Consensus Algorithms in Blockchain for an Efficient and Secure Network." In *Recent Trends and Best Practices in Industry 4.0*, pp. 309–328. River Publishers, 2023.

[10] Tanwar, Sarvesh, Thomas Paul, Kanwarpreet Singh, Mannat Joshi, and Ajay Rana. "Classification and impact of cyber threats in India: A review." In *2020 8th International Conference on Reliability, Infocom Technologies and Optimization (Trends and Future Directions) (ICRITO)*, pp. 129–135. IEEE, 2020.

[11] Gupta, Medini, Sarvesh Tanwar, Tarandeep Kaur Bhatia, Sumit Badotra, and Yu-Chen Hu. "A comparative study on blockchain-based distributed public key infrastructure for IoT applications." *Multimedia Tools and Applications* 83 (2023): 1–26.

[12] Gundaboina, Lokesh, Sumit Badotra, and Sarvesh Tanwar. "Reducing resource and energy consumption in cryptocurrency mining by using both proof-of-stake algorithm and renewable energy." In *2022 International Mobile and Embedded Technology Conference (MECON)*, pp. 605–610. IEEE, 2022.

[13] Prakash, Ravi, V. S. Anoop, and S. Asharaf. "Blockchain technology for cybersecurity: A text mining literature analysis." *International Journal of Information Management Data Insights* 2, no. 2 (2022): 100112.

[14] Dai, Fangfang, Yue Shi, Nan Meng, Liang Wei, and Zhiguo Ye. "From Bitcoin to cybersecurity: A comparative study of blockchain application and security issues." In *2017 4th International Conference on Systems and Informatics (ICSAI)*, pp. 975–979. IEEE, 2017.

[15] Wylde, Vinden, Nisha Rawindaran, John Lawrence, Rushil Balasubramanian, Edmond Prakash, Ambikesh Jayal, Imtiaz Khan, Chaminda Hewage, and Jon Platts. "Cybersecurity, data privacy and blockchain: A review." *SN Computer Science* 3, no. 2 (2022): 127.
[16] Woodfield, Daryl. "The emerging impacts of blockchain technology on DOD asset Cyber Security." PhD diss., Air University, 2019.
[17] Hasanova, Huru, Ui-jun Baek, Mu-gon Shin, Kyunghee Cho, and Myung-Sup Kim. "A survey on blockchain cybersecurity vulnerabilities and possible countermeasures." *International Journal of Network Management* 29, no. 2 (2019): e2060.
[18] Miriam, Hephzibah, D. Doreen, Deepak Dahiya, and C. R. Rene Robin. "Secured cyber security algorithm for healthcare system using blockchain technology." *Intelligent Automation & Soft Computing* 35, no. 2 (2023). DOI: 10.32604/iasc.2023.028850.
[19] Lakhan, Abdullah, Mazin Abed Mohammed, Jan Nedoma, Radek Martinek, Prayag Tiwari, and Neeraj Kumar. "Blockchain-enabled cybersecurity efficient IIOHT cyber-physical system for medical applications." *IEEE Transactions on Network Science and Engineering* 10 (2022): 2466–2479.
[20] He, Songlin, Eric Ficke, Mir Mehedi Ahsan Pritom, Huashan Chen, Qiang Tang, Qian Chen, Marcus Pendleton, Laurent Njilla, and Shouhuai Xu. "Blockchain-based automated and robust cyber security management." *Journal of Parallel and Distributed Computing* 163 (2022): 62–82.
[21] El Bekkali, Abla, Mohamed Essaaidi, and Mohammed Boulmalf. "A blockchain-based architecture and framework for cybersecure smart cities." *IEEE Access* 11 (2023): 76359–76370.
[22] Yu, Sungjin, and Youngho Park. "A robust authentication protocol for wireless medical sensor networks using blockchain and physically unclonable functions." *IEEE Internet of Things Journal* 9, no. 20 (2022): 20214–20228.
[23] Schlatt, Vincent, Tobias Guggenberger, Jonathan Schmid, and Nils Urbach. "Attacking the trust machine: Developing an information systems research agenda for blockchain cybersecurity." *International Journal of Information Management* 68 (2023): 102470.
[24] Fahmi, Novandio, Diki Edwin Hastasakti, Diaz Zaspiagi, Rino Kurniawan Saputra, and Septi Wijayanti. "A comparison of blockchain application and security issues from Bitcoin to cybersecurity." *Blockchain Frontier Technology* 2, no. 2 (2023): 58–65.
[25] Kumar, Adarsh, Deepak Kumar Sharma, and Gaurav Sharma. "Role of cybersecurity and Blockchain in battlefield of things." *Internet Technology Letters* 6 (2023): e406.
[26] Sherin, K.P. Shana, T. Ambikadevi Amma, K. Sivakumar, and S. Divya. "Cyber security in blockchain technology." *International Journal of Research in Engineering, Science and Management* 6, no. 2 (2023): 39–41.
[27] Li, Wenjuan, Yu Wang, and Jin Li. "A blockchain-enabled collaborative intrusion detection framework for SDN-assisted cyber-physical systems." *International Journal of Information Security* 22 (2023): 1–12.
[28] Puthal, Deepak, Nisha Malik, Saraju P. Mohanty, Elias Kougianos, and Chi Yang. The blockchain as a decentralized security framework [future directions]. *IEEE Consumer Electronics Magazine* 7, no. 2 (2018) 18–21.

[29] Saxena, Rashi, and Edamadaka Gayathri. "Cyber threat intelligence challenges: Leveraging blockchain intelligence with possible solution." *Materials Today: Proceedings* 51 (2022): 682–689.

[30] Bansal, Pranshu, Rohit Panchal, Sarthak Bassi, and Amit Kumar. "Blockchain for cybersecurity: A comprehensive survey." In *2020 IEEE 9th International Conference on Communication Systems and Network Technologies (CSNT)*, pp. 260–265. IEEE, 2020.

[31] Hasanova, Huru, Ui-jun Baek, Mu-gon Shin, Kyunghee Cho, and Myung-Sup Kim. "A survey on blockchain cybersecurity vulnerabilities and possible countermeasures." *International Journal of Network Management* 29, no. 2 (2019): e2060.

[32] Alotaibi, Bandar. "Utilizing blockchain to overcome cyber security concerns in the internet of things: A review." *IEEE Sensors Journal* 19, no. 23 (2019): 10953–10971.

[33] Vance, Taylor Rodriguez, and Andrew Vance. "Cybersecurity in the blockchain era: a survey on examining critical infrastructure protection with blockchain-based technology." In *2019 IEEE International Scientific-Practical Conference Problems of Infocommunications, Science and Technology (PIC S&T)*, pp. 107–112. IEEE, 2019.

[34] Rathore, Shailendra, and Jong Hyuk Park. "A blockchain-based deep learning approach for cyber security in next generation industrial cyber-physical systems." *IEEE Transactions on Industrial Informatics* 17, no. 8 (2020): 5522–5532

[35] Singh, Sachchidanand, and Nirmala Singh. "Blockchain: Future of financial and cyber security." In *2016 2nd international conference on contemporary computing and informatics (IC3I)*, pp. 463–467. IEEE, 2016.

[36] Ahram, Tareq, Arman Sargolzaei, Saman Sargolzaei, Jeff Daniels, and Ben Amaba. "Blockchain technology innovations." In *2017 IEEE Technology & Engineering Management Conference (TEMSCON)*, pp. 137–141. IEEE, 2017.

[37] Singh, Narayani, Rahul Patekar, Gaurav Kedia, and Neha Tirpude. "Towards an efficient scheme for user authentication based on decentralized blockchain." *International Journal of Next-Generation Computing* 14, no. 1 (2023). DOI: 10.47164/ijngc.v14i1.1021.

[38] Alajlan, Razan, Norah Alhumam, and Mounir Frikha. "Cybersecurity for blockchain-based IoT systems: a review." *Applied Sciences* 13, no. 13 (2023): 7432.

Chapter 8

# Improving the quality of service of application with Big Data & SDN

*Lipsa Das, Laxmi Ahuja, Deepshikha Bhargava, Bhanu Prakash Lohani, Pradeep Kumar Kushwaha, and Amardeep Gupta*

## 8.1 INTRODUCTION

It is still a challenge to support end-to-end Quality of Service (QoS) in current network topologies. There are still many technical obstacles to overcome in order to manage the quality of real-time multimedia services like networked virtual reality and video streaming. For instance, the demand for data rates from video streaming services is skyrocketing. Low user-to-server latency is required for virtual reality applications at the same time. The growth of the quality concept, which over the last 10 years has evolved from a more technical, network-level QoS to a user-centric Quality of Experience (QoE) [1,2], has led to these conflicting expectations. QoE identifies extra aspects that affect amenities quality as experienced by users, going beyond QoS, which typically communicates network efficiency in the form of quantifiable measures like throughput and delay.

The later QoE influence factors (QoE-IF) may, for instance, comprise the screen resolution of the user device and prior service consumption history. Recent research has implemented QoS-/QoE-oriented network control and management using the paradigm of software-defined networking (SDN) [3] (CaM). The centralised network perspective of worldwide programmability and segregation of the data plane and control plane are its key benefits. Researchers are focusing on these elements to increase the supply of QoS for the various network applications being used today. By altering the network configuration to match the set quality enhancement target, such as lowering the frequency of video stream stalling occurrences, the corresponding "CaM loop" seeks to improve the quality of the service. Architectural building blocks are provided by SDN, which separates network control logic from data plane devices into separate controller entities [4]. Network applications that use SDN can convey data about the statuses of multimedia applications to the network controller on the one hand and to other network applications on the other thanks to well-defined communication interfaces. On the other hand, SDN makes it easier to get data on the controller entities' network-wide performance using tools like OpenFlow [5]. As a result of having access to both stakeholders, SDN is capable of handling an overall application state all over the network.

DOI: 10.1201/9781003504559-8

Due to their exceptionally high volume and level of processing complexity, big data (BD) applications (in particular actual time or near-real-time apps) would be challenging without the underlying assistance of networking [6]. As a brand-new paradigm in networking, SDN is currently attracting a lot of interest [7]. It is now the technology for network management that is most extensively utilised. More and more existing networks are being altered to handle SDN since it enhances the network. The main benefit of SDN is made possible by the network's ability to separate the control and data planes. In a packet-switched network, network routers often construct their own picture of the network topology and attempt to route packets as efficiently as possible using the two-dimensional view of the topology [12]. When the control plane and forwarding plane are segregated, the control plane assumes responsibility for creating the network view, and the forwarding plane merely executes the control plane's instructions. This strategy offers a variety of benefits. The controller, or centralised control plane, has a deeper understanding of the network and can route based on continuously updated data about the network. The controller may alter in response to the time of day or shifting traffic demands. When you consider the current state of internet traffic, this is a significant advantage. Dropbox [14] and Netflix [13] are two examples of hosted services. Dropbox switched from Amazon Web Services (AWS) to its own data centre whereas Netflix remained on AWS [15,16]. Both are well-liked services that, no matter where they are offered, need stable networks to function. The demand for these services, however, is where the disparity lies. While Netflix's demand is significantly higher on the weekends, Dropbox's demand is more consistent throughout the week. According to some estimates, Netflix accounts for one-third of all internet traffic in the U.S. during the weekend. Given that demand is significantly lower during the week, this is substantial. Google and Facebook also account for 70% of traffic in North America [17]. BD is heavily utilised by Facebook [18]. If used, an SDN controller can assist in this situation by altering routing choices based on traffic demand, with AWS managing changes in the number of resources needed to run the service. Many issues could arise during transfer because BD apps were created specifically to manage enormous files over the same networks. The main issue is that because the file is so enormous, the network condition could change throughout the transfer. These issues are connected to how TCP functions and how large files are transported during TCP sessions. Additionally, Hadoop (one such programme that Google Inc. first created to enhance its search. Data nodes are used to store and process files in this scalable large data management system. Within a data centre setting, these data nodes are servers or virtual machines. A file is broken into smaller, easier-to-manage portions before being stored on the system, and each part is stored individually [19]. The file pieces are delivered to the public data network upon request after being retrieved from various locations in the data centre Public Data Network. Clearly, the network conditions in the information centre have an impact on the portion of transfer that takes place there. SDN can assist in achieving greater efficiency in this area [20].

So, developing and deploying applications in SDN is simpler than it is in traditional networks. Additionally, enforcing the consistency of network policies is simple using SDN's global perspective. SDN brings a novel speed of advancements in networking architecture and marks a significant paradigm shift in the network's progression.

Although there has been some good work on BD and SDN, these two crucial issues have often been covered separately in the majority of earlier studies. On the other hand, SDN will have a huge impact on BD applications as a key networking paradigm. Several advantageous features, like a division of the control and data planes, conceptually centralised control, a global view of the network, and the ability to programme the network, can make BD collecting, transmission, storage, and computation considerably simpler. As an illustration, massive data processing frequently occurs in data centers in the cloud centres.

SDN-based data centres can outperform conventional data centres by dynamically allocating resources in data centres to different BD applications to achieve their service level agreements [8,9].

BD, a crucial network application, will have a substantial impact on the creation and execution of SDN. In particular, given the global view of the network, the conceptually centralised controllers in SDN can gather enormous amounts of data at any level of detail from all the various layers (from the physical to the application layers). We've learnt from previous cross-layer design projects that while exchanging information between layers can enhance network performance, doing so can make the network so complicated that conventional design and optimisation techniques are ineffective. Fortunately, the design and management of SDN can benefit from BD analytics, which uses analytical techniques to extract insights from data to inform decisions. For instance, using large traffic data analytics to improve SDN performance is easier for the controller to implement [10].

Like massive amounts of data can be processed effectively thanks to recent advancements in BD technology, like Deep Learning or MapReduce. Additionally, the majority of the most cutting-edge machine learning (ML) and data mining (DM) algorithms may be executed efficiently thanks to BD approaches. Integrating SDN control logic and BD analytics techniques, such as by incorporating them into an SDN controller, would make it possible to take into consideration a variety of QoE-IFs and, as a result, make "more precise" decisions that are in line with the defined CaM aim. Additionally, BD approaches might be applied, for instance, to tailor QoE estimation models during service run-time to take into account user categories with various demographics.

The article discusses the fundamental qualities and contemporary developments of BD and SDN and then gives a summary of the most recent findings in the areas of QoS or QoE CaM and methodologies for QoS or

QoE CaM for real-time multimedia amenities enabled by SDN. The pros of integrating BD analytics with CaM based on SDN for the aim of enhancing or optimising QoS or QoE are then demonstrated using three particular use-case situations related to video streaming amenities.

## 8.2 OUTLINE OF SOFTWARE-DEFINED NETWORKING ALONG WITH BIG DATA

We go through the fundamental ideas and characteristics of BD and SDN in this section, as well as their brief interactions.

### 8.2.1 SDN (software-defined networking)

Due to the increasing amount of data from various applications on the Internet, SDN has been viewed as one of the most likely methods to change the cyber world with better network management [7]. As the first SDN protocol to be established, OpenFlow is sometimes referred to as SDN's enabler. At Stanford University, the phrase "SDN" was initially used to describe the concepts and work done with OpenFlow. In accordance with the original definition, SDN alludes to a network infrastructure where a remote control plane that is disconnected from the first manages the forwarding state in the data plane. Standards for SDN protocols are being created in live applications at an increasing rate. These days, the term "SDN" refers to the platform as a whole and is not specific to any one implementation [10].

As stated by the Open Networking Foundation, an organisation that is not-for-profit dedicated to the growth and standardisation of SDN, the control and data planes are separated, network understanding and state are effectively centrally controlled, and the network's fundamental infrastructure is segregated from the apps in the SDN architecture. For SDN, four key elements can be found:

1. A decoupling between the control and data planes.
2. Rather than destination-based, forwarding decisions are flow-based.
3. The so-called SDN controller, also known as the network operating system, receives the centralised control logic (NOS).
4. Software programmes that interact with the underlying data plane devices and run on top of the NOS can be used to programme the network.

These beneficial characteristics can make acquiring, transmitting, storing, and processing massive data much easier.

In its reference architecture, SDN can be broadly separated into three layers: the infrastructure layer, the control layer, and the application layer. Different SDN applications can be seen at the application layer of the SDN architecture. Through the northern app programming interfaces, the application layer communicates via the SDN control plane regarding the condition of the network and its particular requirements. The main element of the SDN model, which regulates how the network as a whole operates, is thought to be the SDN control plane in the control layer. Since NOX, the original SDN controller, a number of other SDN controllers have been created, notably Onix, POX, Beacon, Ryu, and Floodlight. Software-defined network has lately been used in the topologies of data centre networks to improve performance with minimal complexity and energy consumption [8]. SDN may dynamically assign resources to increase the efficiency of applications by gathering traffic data. Additionally, a recent development is the use of SDN's capabilities for application-aware networking in network topologies and applications for data centres.

### 8.2.2 Big Data

Recently, BD has gained popularity as a research area. The capability to keep and execute information sets has a significant impact on the history of BD. Innovative technologies are created to keep and handle ever-bigger information sets, ranging from megabytes-gigabytes-terabytes-petabytes-exabytes, as capability has increased over time. In the literature, big data is defined in a variety of ways [11]. BD is described in terms of five key characteristics in the attributive definition: amount, variety, speed, significance, and veracity. BD is described as "data sets of a size beyond the capability of typical database tools to gather, store, handle, and analyse" in the comparative definition. BD's generating domains are connected to its sources. Business (such as business-to-business and business-to-consumer transactions), networking (such as the Internet, cellular networks, and the Internet of Things [IoT]), and scientific studies (such as biological computation, astronomy, and high-energy physics) are a few examples of the generating domains.

Enterprise applications, IoT-based apps, applications for healthcare, and scientific research applications are just a few examples of the many areas where BD may be highly helpful. The majority of the market leaders, such as Facebook, Amazon, Google, Microsoft, and Oracle, have started to build BD projects. The National Science Foundation, the Defense Advanced Research Projects Agency (DARPA), and the National Institutes of Health have all made significant investments in BD research [10].

## 8.3 CATEGORISATION OF CONTEXT-AWARE NETWORKING TECHNIQUES BASED ON SDN

The next part discusses several QoS or QoE-focused network control and management abbreviated as QoS/QoE CaM strategies that make use of

SDN. We categorise the demonstrated techniques or approaches or methods that are QoE-based influence factors abbreviated as QoE-IFs, the control activities that are prompted by this information, and the implications that follow, as, in specific, information variety, data gleaned from monitoring, data analytics, and associated control actions plays a major part in the BD context. The summary of the explored methodologies along with a classification is shown in Table 8.1. The following solutions all revolve around video streaming or video conferencing as an outcome of this, as video streaming currently makes it a great deal of Internet congestion [21] and serves an important part with representatives like MPEG DASH and HTTP adaptive streaming results.

### 8.3.1 QoE influence factor monitoring (QoE-IF)

We start by categorising chosen management strategies in accordance with the factors under observation. Each SDN-based CaM strategy keeps an eye on at least one QoEIF. The number and location of monitoring can be

*Table 8.1* Categorisation of context-aware networking techniques based on SDN

| *Source* | *Monitored QoE-IFs* | *Control Action* | *Implication* |
|---|---|---|---|
| [8] | Packets in the network (DPI) | App-aware path selection | Prevention of video stallings |
| [9] | Packet loss, delay | App-aware path selection | Quality of Experience (QoE) enhancement of a multitude of applications |
| [10] | Network congestion indication | Flow prioritisation | Quality enhancement of live video transmission |
| [11] | YouTube Video Buffer | App-aware path selection | Prevention of video stallings |
| [12] | Network throughput, video buffer | Flow prioritisation and quality adaptation | Prevention of video stallings |
| [13] | Active DASH streams, available bandwidth, client properties | Bandwidth reservation | Prevention of video stallings |
| [14] | Available bandwidth, network latency, client properties | Dynamic resource allocation | Fair QoE maximisation |
| [15] | Available bandwidth, packet loss rate, jitter, initial delay, buffer | Change of routing paths and transport nodes | QoE enhancement for video streaming |
| [16] | Active DASH streams, network resources, client properties | Bitrate guidance, bandwidth reservation | Fairness with respect to video quality |
| [17] | Available bandwidth, video buffer | Network resource allocation | Fair QoE maximisation |

used to categorise these factors. We take into account the four monitoring dimensions listed below. There are typically mechanisms that perform

i. network tracing at the packet level,
ii. network tracing at the flow level
iii. network tracing of application data that is accessible in the user software,
iv. network and application-side tracing, that is, network parameter monitoring as well as application-side QoE-IFs updates.

#### 8.3.1.1 Tracing at packet level in the network

QoE management of web and video traffic, [22] by Jarschel et al. it makes use of network information at the packet level. Deep Packet Inspection (DPI) is the process of inspecting packets as they travel from source to destination. The application can be recognised based on important packet fields. The difficulty is that in order to obtain a comprehensive picture of the application in the network, a variety of information must be gathered at various locations in the network. It also addresses whether end-to-end encryption of encrypted data creates another significant challenge, raising the question of whether northbound interaction between an application and an SDN controller is preferable to packet DPI. The method from [23] gathers data on packet delay and loss while forgoing extracting application information from packet payloads. The collecting of packet-level information at various network nodes presents a significant barrier in this situation. The ability to monitor the network as a whole has a significant impact on the effectiveness of these methods and the accompanying determination of network bottlenecks.

#### 8.3.1.2 Tracing at flow level in the network

There are methods that investigate the network at the flow level in addition to those that monitor it at the packet level. Individual packets are not examined for this reason; instead, network flow is taken into account. According to network flow statistics, monitoring is carried out in [24], for instance, to immediately identify network congestion.

#### 8.3.1.3 Tracing of application information

Reference [25] is an illustration of how application data is utilised as the foundation for control actions. The user's buffer state is used as the basis for this proposal's QoE-IF. The client buffer state is just one of several examples examined in this case for the use case of video streaming. Greater information collection at the application level may reveal that targeted monitoring of certain application parameters is required for every running program and for every customer in order to optimise QoS or QoE management in the network.

#### *8.3.1.4 Tracing at network- and application-side*

The techniques provided in [26–31] take into account both application and network information for determining control actions, compared to the solutions stated before, which solely depend on network surveillance or application tracking. This may comprise, among other things, the quantity of running video streams, the screen resolution capabilities of the devices being utilised, and QoE-IFs like the current buffer or the quantity of quality switches. As a result, gathering extensive data on the traffic patterns of other apps in multi-application environments presents a new issue. While doing so necessitates monitoring a wide range of data, it also has the advantage of allowing decisions to be made that will benefit all applications. Given that data volume and variety are essential to the effectiveness of BD and machine-based data analysis, all of the aforementioned problems are directed in that direction.

### 8.3.2 Techniques of control actions of management

The offered procedures rely on numerous network changes to fulfil an application's requirements. Some systems also consider extra control actions in the application and also the control actions executed in the network. Since SDN is the main focus of our work and automatically handles the network-side adaptations using the OpenFlow Southbound API, we will only go into depth about the network-side adaptations in the sections that follow. Many of the suggestions use the same or comparable control actions, as the classification table that is supplied demonstrates. In [22,23,25], and [29], dynamic application-aware path selection is carried out, and in the latter, a dynamic selection of the transport node is also carried out. An algorithm chooses the network path for the particular flows in order to satisfy the application needs using information from monitoring entities and QoE models. Making decisions in this situation that are coordinated and precise is difficult. Therefore, the decision-making process is enhanced by the information's granularity. The network can be utilised to perform more specific operations using fine-grained information. It is feasible to more accurately estimates the necessary application components affecting the control operations, like users' video buffers, if ML is additionally applied to the mass of data. The volume of data is crucial for all methods in order to make effective decisions without prejudice towards other network applications.

Specific network flows are momentarily given priority by the methods of [24,26] so as to stop QoE degradation. It is accomplished by setting up a minimum of two queues, out of which one is configured as a best-effort queue, whereas the other one handles the packets from the prioritised flows. As opposed to the packets in the best-effort queue, the ones in the high-priority queue are preferred to be scheduled.

In [27,28,30,31], dynamic resource allocation, such as bandwidth reservation, is taken into consideration. The fact that these algorithms include fairness factors unites them. This might either imply that all video clients have an equal video quality despite the fact that they may have various device capabilities (such as screen resolution) and, consequently, varying demands on the network, or that the QoE is maximised while taking into account fairness limitations.

These mechanisms in the decision-making process may be supported by BD and ML strategies. It is possible, for instance, to employ strengthened learning. An algorithm picks up new information from the effects that certain control actions have. As a result, it continuously seeks to optimise its choices and is aware of the control action that is currently most appropriate. The comprehensive feature set taken into account for learning, which serves as the algorithm's decision-making foundation, includes, among other things, various things, typically the behaviour of networks and applications. It is even feasible to respond in a user-centric way by leveraging user-defined data gathered from monitoring user behaviour (for example, video-stalling length or initial delay thresholds that cause the user to abort).

### 8.3.3 Big data potential for managing SDN QoE

As was covered in the last sections, it is undeniable that SDN-based QoE management systems could benefit from the use of BD. The usage of BD in this area is mandated by the trend towards greater data and more closely watched QoE-IFs. However, current ways fail to take advantage of this capability and prevent using big data since using this volume of data requires a different strategy and way of thinking. Big data assists in assessing information in three separate directions, in contrast to traditional methods. The statistical analysis of encrypted traffic is supported by big data, which is crucial in today's networks. Traditional methods avoid packet analysis due to privacy concerns and instead gather statistics on network and packet throughput to learn more about the applications. Second, big data can aid in the analysis of data throughout the entire network at various points of presence. By considering data about the entire network, many CaM techniques can enhance their optimisations. Thirdly, BD is useful at the application level where many pieces of data must be gathered and every application must be taken into account.

BD can assist with its analytics techniques in addition to the gathering and analysis of the massive volume of data. Algorithms that help grasp and forecast relevant control actions that are based on the provided information, for instance, are feasible. In Section 8.1.3.2, some recent developments in the field of BD and QoS/QoE management are discussed, along with instances in which network control decisions are based on the results of BD analytics techniques.

Making coordinated, precise decisions is difficult when dealing with control actions. In order to make effective decisions and avoid treating other apps on the network unfairly, sufficient data must be available for all techniques.

## 8.4 BIG DATA ANALYTICS SUPPORTING MANAGEMENT OF QOS OR QOE

A brief review of typical BD analytics methodologies is provided in this subsection. We won't talk about BD in general because it handles a wide range of data handling issues in addition to ML and DM techniques [32]. Aspects of data management are only marginally significant in this context because we must cope with the "3 Vs": Volume: We must manage enormous amounts of information, Velocity: We must use incoming data quickly, and Variety: We must reduce the gap between the network flow data and application level. We disregard the additional "2 Vs" of BD. Thus, we would concentrate on common ML and DM techniques that serve as the foundation for an examination of the gathered data with a focus on BD characteristics. Reference [33] has a detailed explanation of the integration of BD and SDN.

ML's objective is to create models by learning from a set of instances. This model can later be used with brand-new, unexplored data.

The second objective is to use those models to discover new information about the current data.

This model-learning step is a part of the larger process of DM, which also involves processing the data and using the learned models. Numerous innovative "DM" methods have been created in the past as a result of this data-centric viewpoint. The association rule mining method, which belongs to a larger class of techniques called pattern mining, is the most well-known example. In general, there exist three types of ML and DM methods: supervised learning, semi-supervised learning, and unsupervised learning. References [34–36] and [37–39] provide an introduction to ML and DM, respectively.

The classification rules (model) are trained based on a labelled dataset when supervised learning is used. Depending on the given input, labelled data, also known as the training dataset, shows the desired outcome, or the right feature value. As a result, the model creates a function or relationship that links the input parameters with the outcome features. The outcome is then projected based on the model's application to unlabelled data. Support Vector Machine, Decision Tree, Naive Bayes, k-Nearest Neighbor, and Random Forest are examples of common supervised learning algorithms. These methods frequently fall under the classification umbrella. Regression forecasts continuous values as opposed to classification, which produces output variables that belong to a predetermined class. Other instructional strategies, such as bagging, boosting, or ensemble learning, integrate both strong and weak graspers to create a new model.

Clustering refers to unsupervised learning techniques in which no pre-labelled data is provided. Expectation-Maximization, basic statistical methods like k-Means, and density-based techniques like DBSCAN are examples of common clustering techniques (For a survey [40]). LDA-style techniques have gained popularity in numerous fields more recently. Textbooks like [36] already cover the fundamentals [47].

Unsupervised learning methods without the use of labelled data are referred to as clustering. Common clustering approaches include Expectation-Maximization, fundamental statistical techniques like k-Means, and density-based algorithms like DBSCAN. (For a survey [40]). Recently, LDA-style approaches have become more common in a variety of domains. The essentials are previously covered in textbooks like [36].

The model learning or pattern discovery stages of the previously stated ML and DM methodologies often require access to all data.

Sometimes it is hard to save all of the "Big" data, making it impossible to use traditional ML and DM techniques. The term "stream DM" refers to a group of techniques used so that models may be discovered or patterns can be immediately recognised from a stream of data. In addition to widely used conventional methods like tree learner, there are specialised approaches in this field such as time series analysis that depend on data streams by nature. Reference [41] provides a summary of the stream mining algorithm. We can address the BD issue; since, storage of data is no longer required, we are seeing a volume and velocity data issue in our network analysis setting. One solution is to transmit data directly to a machine that learns and builds new models as needed. These models can be installed on network controllers or devices, and they manage network flow. To effectively learn from a data stream, one can use the distributed stream processing framework offered by the Apache Storm framework 1.

Other ML paradigms that allow an agent to autonomously learn a tactic are reinforcement learning (cf. [28]). It could be considered a form of ineffective supervision because there is little feedback given to help the worker understand the technique. The agent is not taught what to do; rather, it is rewarded for its choices, both positively and adversely. The feedback from playing games, such as winning or achieving a higher score, or operating machinery, like a robot navigating a maze, are typical examples. The advantage of such a system is that it continuously learns, even when it is being used in real-world situations.

As demonstrated by instances like game learning for Atari computer games [42], reinforcement is growing in popularity and success as deep learning technologies progress. The fundamental concept is to carry out the various processing stages on the corresponding layer using a deep neural network. This covers both the processing of images, which in a traditional learning phase would be a form of feature engineering, and the evaluation of the reward over an extended period of time. Without getting into the specifics of how deep learning algorithms operate, one may use comparable deep learning techniques to analyse network traffic in order to regulate how the traffic moves through the network. By directly examining the network traffic, the neural network could grasp the network reconfiguration. By utilising one of the stream frameworks, this might likewise be accomplished in a stream setting.

If stream mining is not possible, BD must be stored and processed for learning. There have been many common paradigms that have evolved in recent years. One of them is the MapReduce methodology used in Spark (memory focused) or Hadoop (disk focused), which was primarily created by search engine companies to execute web-scale information. Large datasets can be processed quickly and effectively with MapReduce [43]. The key of the technique is formed by the two functions map and reduce, which iteratively call one another.

While the map phase, the input information is organised in accordance with specific requirements represented as a client-defined function. The reducers distribute the results after receiving them. In order to produce fewer results or perhaps only the end outcome set and return it, the reductions condense the data. If the information is arranged in a distributed manner, the initial map activity will be given immediate access to it, allowing us to work on a large amount of information simultaneously and fast. NoSQL databases are a paradigm that has recently been established to store enormous amounts of data.

These databases store data using different primary principles than traditional relational databases, like columns, documents, key-values, graphs, and many models. In addition to changing the storage model, these databases prioritise speed over more conventional qualities like consistency. Reference [32] provides an overview of the emerging, frequently distributed storage models.

## 8.5 INTEGRATING BIG DATA ANALYTICS AND (SDN) SOFTWARE-DEFINED NETWORKING: 3 USE CASES TO BOOST QOS OR QOE

Three use cases that demonstrate the anticipated advantages of merging BD with SDN are shown in this section. Expansion of traditional network QoS monitoring is the first use case to accomplish network enhancements and adjustments as a result of specific network configurations. In the second use case, BD-applications are used to process the values of QoE-IFs that are exchanged among a video instant on demand contributor and a network operator based on SDN. Unlike the second use case, which requires direct interaction between the video amenities provider and the network operator, the third use case uses BD applications to infer amenities-level QoS or QoE.

### 8.5.1 Improving network performance

This section focuses on the use case of enhancing network performance through the integration of BD with SDN. We specifically talk about how the network performance complies with VoIP traffic's QoS criteria.

In today's networks, traffic flows and their interactions with one another are exceedingly complicated and unpredictable. Network-level operations like queuing, congestion shaping, selected dropping, and link-effective strategies gives a network operators control over how the flows transition through the network. It is crucial for VoIP and streaming video congestion as the operator must ameliorate the network's efficiency and uphold the established QoS criteria, like the greatest permitted latency and the minimal necessary throughput. Technically speaking, this means that the network settings and QoS optimisations can be employed to maintain a robust traffic flow in situations where network virtualisation isn't feasible or where the usage of innovations like virtual local area networks is insufficient.

VoIP networks are a prime illustration. There, the one-way delay is typically limited to 400 ms [44] for telephony applications. If VoIP traffic is being transmitted, then this must be true across the board. Layer 2 prioritisation with Class of Service markings or layer 3 markings, ideally Differentiated Services Code Point, is usually used in this case in the outbound direction of every network link.

The important network quality aspects are continuously measured and monitored as the foundation of VoIP QoS management. Currently, the network uses switches and routers to generate NetFlow statistics on the delay of packets and to carry out active tests of how to meet the QoS requirements for VoIP. In addition to the traditional monitoring and testing, two basic concepts can be used with BD and BD Analytics: (1) using BD processes to gather, archive, and analyse information at an advanced level of detail, and (2) the examination and assessment of the data with BD methods of learning to deliver greater understanding, identify errors, anticipate the future critical circumstances, and use trends without direct operator interaction.

#### *8.5.1.1 Data collection, archival, and execution in accordance with big data principles*

The complete network's QoS statistics can be gathered through the mass collection and storing of data. To perform better network control, it is also feasible to incorporate application information (Variety) as an extra source.

BD offers tools for effective data storage, such as NoSQL databases, the scaling of storage clusters that is based on information volume, and the processing of information to support analytical engines like Hadoop. Due to the larger volume of the new data, more precise statistics are possible. Furthermore, these data serve as the foundation for creating new models and uncovering the network's underlying usage pattern knowledge.

The new size has resulted in more precise insights and actions that can be taken in response to user needs more quickly (velocity). When learning is employed, this might potentially develop to the point where the network can automatically handle requests for personalised traffic.

#### 8.5.1.2 Using big data learning techniques for analysis and evaluation

By studying the gathered information and deriving QoS models from it, conclusions on QoS-compliance could be drawn out. The enforceability of the QoS standards and the requirement for QoS modification in light of models built using historical data can both be determined. Patterns of day-to-day and congestion problems could be predicted, managed effectively, and assessed for network control goals. Using the collected data, it is also feasible to make long-term modifications and forecast future traffic conditions. In Section 8.1.3.2, instances of effective ML applications on network traffic are presented to illustrate what is now feasible with state-of-the-art models. We envision self-adaptable SDN controllers with the use of deep learning and stream DM techniques and a high-level network provider approach that showcases the full promise of BD analytics in this region.

With the aid of SDN, QoS management is implemented and configured throughout the entire network. For instance, using SDN, control commands are sent to the gadgets, and dynamic changes could be performed on the basis results of the Big Data-analytics engine.

The application of BD to QoS management is the logical extension of the concept whereby information is analysed to establish QoS needs for specific kinds of multimedia amenities.

### 8.5.2 Using business agreements to improve the calibre of streaming video on demand

Another illustration of how the integration of BD-applications is provided in this subsection potentially improves QoS/QoE in a network environment powered by SDN. The example presupposes that an upcoming SDN-based network operator (SNO) and a streaming video (Video on Demand [VoD]) streaming facility provider (SP), like Netflix or Amazon Prime, have achieved an understanding to interchange network- and service-level data important to QoS or QoE regulation.

These business arrangements between SPs and SNOs may be advantageous to both parties: SPs provide enhanced QoS or QoE to their customers, while NPs can make better use of their network resources. The business accord covers a variety of topics. The SNOs consent to give the SPs' clients "prioritised" traffic treatment.

Additionally, SPs and SNOs concur on the value interchange for the pertinent QoS or QoE influence factors.

We discovered the following criteria for VoD streaming that the SPs must report:

- (Anonymised) user demographic information that is reported at the establishment of the video session, such as the user's age range;
- prior service usage history (novice/expert user), which is disclosed at the video session setup step;
- cost of the service (fixed charge, cost per video, etc.), which is disclosed during the setup of the video session.
- the client's gadgets type (such as a laptop, tablet, or smartphone), which is reported at the establishment of the video session;
- client gadgets features, which are reported during the setup of the video session (size of screen, OS, Central Processing Unit and RAM characteristics, etc.);
- video client statistics, which are periodically reported for the duration of the session (e.g., buffer status, quantity of video hanging);
- service characteristics (MPD data), which are outlined throughout the video setup phase.
- for the duration of the session, server statistics are released on a regular basis.

However, there are a lot of QoE-IFs in this, and SPs must effectively deliver and monitor them. Here, the SPs may use BD-applications to quickly analyse and minimize the traced information on clients and VoD amenities.

Additional implementation steps are required to put such an architecture into operation. First, the QoS/QoE-IF reporting would be added to the VoD clients and servers, perhaps by piggybacking on HTTP congestion. For the data interchange between SPs and SNOs, an orchestrator that is considered the gathering information point and retrieval for information on relevant influence factors may be used. To connect to the SDN control plane, the orchestrator can make use of a northbound interface offered by one of the freely available SDN regulators.

There are two approaches to utilise BD services. Either the orchestrator integrates Big Data applications directly or it uses another interface to communicate with a BD infrastructure. Additionally, the orchestrator's performance may be enhanced during runtime with the aid of BD apps. For example, it may be possible to extract delay information, which enhances network optimisation.

Additionally, BD-applications free SDN controllers entirely of processing raw data, easing the load of intensive information processing on them.

The efficiency of the data plane elements across the entire network may be periodically compiled by SDN controllers on the SNO side. Based on the knowledge obtained via the orchestrator and the network surveillance information, the control plane can make decisions in the most optimal manner consistent with the business accord and the general objective of network optimisation. The VoD service and end-user data could be used by the SDN controller to guide decision-making. A beginner is less likely to find visual flickers annoying than an advanced user, so the SDN controller may be able to distinguish between advanced users and beginners using this information. The price of the service could be another piece of end-user data. In this case, a user who pays for every video anticipates greater value for their money than a client who pays a flat cost. BD apps running on the SP side can give these metrics and insights. Additionally, additional BD-received data can shed light on the causes of video freezes. In this case, end-user data makes it possible to distinguish between video freezes caused by improperly configured network activity and video freezes caused by poor client efficiency, like an overloaded gadget running too many applications. Frequent variations in video quality can provide additional service information. As a result, video traffic can be supported by the SDN controller to offer a highly reliable delivery and hence minimise quality alteration. In such situations of general over-usage, data received from BD apps enable the identification of consumers depending on their prior use patterns or history of programme usage.

### 8.5.3 Enhancing applications without business agreements in quality

BD applications can nonetheless contribute to enhancing the general performance of the service and the network in a situation where there is no direct negotiation and data sharing between SNOs and SPs. An SNO has an incentive to provide end customers with the best network performance in this situation because users typically hold SNOs accountable for poor service quality.

Finding the causes of service performance declines, particularly for encrypted network data, is a difficulty. Therefore, the SNO's aim would be a network supervision architecture with an SDN-based regulator that takes decisions according to effectively observed information. The monitoring infrastructure may use BD approaches to relieve the controller of the strain of extensive data processing. Then, based on variables like mean packet dimensions, inter-arrival packet time, and mean throughput, BD-applications would offer statistics of the media stream congestion.

It would be crucial to efficiently convey low-dimensional data because there may be a significant amount of data in this case as well. BD unsupervised approaches or autoencoders could do this. The direct connection and supervising of SDN infrastructures via OpenFlow could be supplemented by other methods, like Simple Network Management Protocol or sFlow.

Such a method, nevertheless, requires models that can determine the values of the pertinent QoS or QoE metrics solely from the traffic features that are observed [45], as video content may be encrypted. Future BD-based network optimisation may potentially contain user research and automatically created QoS/QoE models, even if these models are currently developed via human subject testing. Furthermore, test participants who utilise end-user supervising tools or even directly offer response on QoS or QoE could provide information about end-user usage of the service. Massive amounts of data would be produced by these kinds of systems, necessitating the effective processing of a BD infrastructure. The SDN controllers would use such models, which would be updated in real-time.

## 8.6 AUTO-OPTIMISATION FOR THE INTELLIGENT NETWORK

A true BD application would make use of the thousands of events and transactions that occur every second in an operator's network with the aim of maximising quality and efficiency. Thanks to improvements in analytics and the flexibility of the cloud, it is now possible to manage this massive amount of data and get rapid insight into the condition of the entire global network. The information contained in the whole network at any one time within a specified period of time is accumulated to generate the global network state. In particular, current BD technology trends facilitate such a distributed state gathering and efficient processing, like systems for networked mining of data and data retrieval, such as Spark [47] and Hadoop [46].

To create a telco cloud system, distributed sites are added and connected via SDN-based networks, allowing for an automatic response to both local and global information.

A high-level overview of the interactions between a distributed telco cloud network and a big data analytics engine is shown in Figure 8.1. An illustration of a telecom cloud can consist of three levels, or micro points of presence (micro-pops), central offices, and centralised information centres.

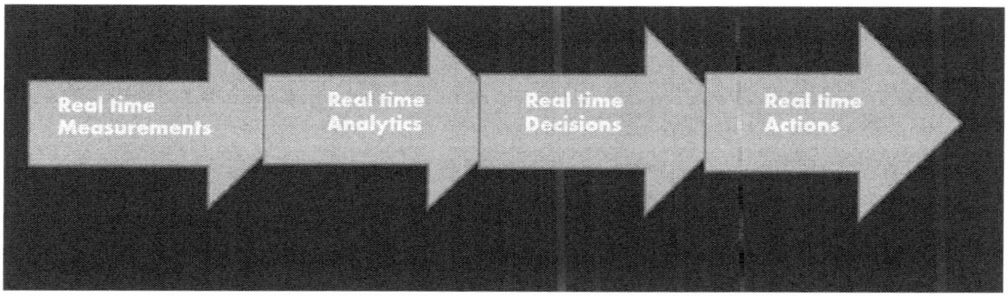


*Figure 8.1* Steps for local network optimisation with huge data.

The network performance for telco cloud services is offered by virtual network functions that are executing within the data centres of every level. The access networks link micro-populations and central workplace, whereas the backbone networks connect central offices to central information centres. The infrastructure apes the Future X Network's [43] description of the global-local based cloud. The primary components of each instance in the three levels are storage, computation, and resources for networking. While large data centres may contain hundreds of servers together with the appropriate storage and networking devices, a micro-pop may just have a few servers and a small storage system linked over a limited SDN network. This is because each category varies in size and number. As a result, the network has a huge number of micro-populations but few major data centres. The local orchestrator and SDN controller for each instance of each tier is displayed elaborately for the centralised office.

Major cities are often where one may find the central office, which is made up of numerous racks. The few hyper-scale data centres have high-speed optical networks connecting each of their data centre locations. The actual time optimisation of the consumer's experience during a single session necessitates the following extra phases among all tiers, regardless the simple fact that this framework permits extensive bandwidth and agile movement of virtual procedures among sites (cf. Figure 8.1). These steps consist of actual time evaluations, real-time analysis, immediate choices, and real-time actions.

The exchange of information between tiers is very important to enable a worldwide optimisation of the network's operation. With the expanding processing capacity of BD applications, the cost of transporting every piece of data would be prohibitive. As a result, comprehensive and condensed representations are necessary, and BD applications that first process the information locally at each layer could meet this need. Furthermore, using such a paradigm, BD-applications could effectively discover latent variables of the tracked networks, or state information. The telco cloud makes this possible because all tasks can be carried out in any of the nearby data centres. The local analytics engine receives data from the regional surveillance system and produces a suggested course of action for the local orchestrator. The orchestrator therefore has two duties. According to the analytics results, the deployment of the pertinent virtual functions in its domain is optimised, and on the contrary, the SDN controller provides directions to route network traffic accordingly.

The local analytics engine not only does local optimisation at every tier, but it also recognises and compresses data that is globally pertinent to the network. The global orchestrator is responsible for collecting the necessary data from the core data centres and sending it to the global analytics engine. The type of data collected depends on the preferences and optimisation objectives set by the global orchestrator. For example, if the goal is to optimise for cost reduction, the global orchestrator might collect data regarding resource usage and pricing. If the goal is to improve system performance, the global orchestrator might collect data related to network latency and system load. Once the data is collected, the global orchestrator will process it and provide the global analytics engine with the relevant information. The frequency of data collection and processing depends on the preferences and optimisation objectives. For example, if the goal is to optimise for cost reduction, the data might be collected and processed every hour. If the goal is to improve system performance, the data might be collected and processed more frequently. Together, they optimise the entire network using larger timescales and macroscale trends than the local metrics.

Thanks to the pre-processing and selection performed at local locations, the network is not clogged with supervising data; rather, the worldwide engine only requires coping with data that is genuinely relevant. An unresolved research issue is the appropriate level of information granularity required for such a global optimisation system. However, if the appropriate abstraction can be identified, an operator can enable basic network modifications, such as core network reconfiguration, as well as automated forecasting of needed modifications and infrastructure faults. By doing so, the operator can reduce operating costs and human error associated with network settings. Together, they optimise the entire network using larger timescales and macroscale trends than the local metrics (Figure 8.2).

The network is not clogged with supervising information thanks to the pre-processing and selection done at local locations; instead, the global engine only needs data that is actually pertinent. An unresolved research issue is the appropriate level of information granularity required for such a

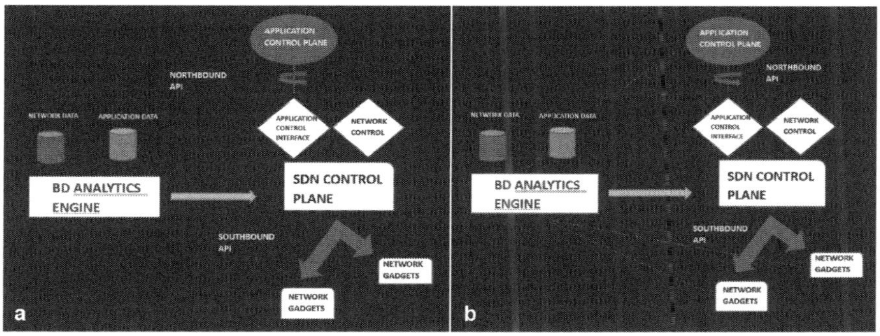


*Figure 8.2* Several infrastructural alternatives for a Big Data/SDN structure.

(a) The big data analytics engine interacts with the SDN controller from a controller-centred perspective.
(b) The big data analytics engine's and the SDN controller's information-focused communication.

global optimisation system. However, if the appropriate abstraction can be identified, an operator can enable basic network modifications, such as core network reconfiguration, and automated foretelling of required changes to and failures of the physical architecture. By doing so, the operator can reduce operating costs and human error associated with network settings.

## 8.7 CONCLUSION

Networks nowadays must deal with a wider range of services that are more demanding than before. Applications span from low-latency industrial applications to high-bandwidth multimedia applications and several IoT services. These various demands and the rising user base necessitate more effective network resource management. Applications' context information is needed for this, as well as a networking infrastructure that can execute resource control operations.

In order to extract such context information, for example by reducing QoS or QoE mappings and to upgrade it on the basis on real information, BD approaches are viable options. This is due to the requirement to correlate enormous amounts of monitoring data from networks and applications. An SDN controller in an operator's network can use this data to improve the quality of the application for certain users.

The research contributes to a greater comprehension of how SDN designs and BD approaches can be combined to ameliorate the overall quality of an application. As a result, we presented a number of SDN-based methods that enhance client QoE by tracking QoE-IFs and taking the necessary network or application control actions. We then emphasised the capability of BD Analytics to enhance QoS or QoE management by presenting numerous works that make use of ML methods in the context of QoE. We expanded on this idea to envision how networks will be able to optimise themselves in the future with the use of BD and ML techniques. At last, problems and unresolved difficulties related to a SDN framework that employs BD to improve the applications' quality were also shown.

## REFERENCES

[1] Qualinet White Paper on Definitions of Quality of Experience (2012); 2013. European Network on Quality of Experience in Multimedia Systems and Services (COST Action IC 1003), Patrick Le Callet, Sebastian Moller and Andrew Perkis, eds., Lausanne, Switzerland, version 1.2.

[2] Varela M, Skorin-Kapov L, Moor KD, Reichl P, Qo E. Defining a User-centric Concept for Service Quality. In: Chen CW, Chatzimisios P, Dagiuklas T, Atzori L, editors. *Multimedia Quality of Experience (QoE): Current Status and Future Requirements*. John Wiley & Sons, Ltd; 2015. pp. 5–27.

[3] Kreutz D, Ramos FMV, Verssimo PE, Rothenberg CE, Azodolmolky S, Uhlig S. Software-Defined Networking: A Comprehensive Survey. *Proceedings of the IEEE*. 2015;103(1):14–76.
[4] Schatz R, Fiedler M, Skorin-Kapov L. QoE – Based Network and Application Management. In: Moller S, Raake A, editors. *Quality of Experience: Advanced Concepts, Applications and Methods*. Springer International Publishing; 2014. pp. 411–426.
[5] McKeown N, Anderson T, Balakrishnan H, Parulkar G, Peterson L, Rexford J, et al. OpenFlow: Enabling Innovation in Campus Networks. *SIGCOMM Computer Communication Review*. 2008;38(2):69–74.
[6] Yu S, Lin X, Misic J. Networking for Big Data. *IEEE Network*. 2014;28(4):4.
[7] Kreutz D, Ramos FMV, Veríssimo PE, Rothenberg CE, Azodolmolky S, and Uhlig S. Software-Defined Networking: A Comprehensive Survey. *Proceedings of the IEEE*. 2015;103(1),14–76. doi: 10.1109/JPROC.2014.2371999
[8] Hong W, Wang K, Hsu Y-H. Application-Aware Resource Allocation for SDN-Based Cloud Datacenters. *Proc. Int'l. Conf. Cloud Computing and Big Data 2013*, Fuzhou, China, Dec. 2013.
[9] Han Y, Seo SS, Li J, Hyun J, Yoo JH, Hong JWK. Software Defined Networking-Based Traffic Engineering for Data Center Networks. *Proc. 16th Asia-Pacific Network Operations and Management Symp.*, pp. 1–6. Taiwan, Sept. 2014. doi: 10.1109/APNOMS.2014.6996601
[10] Cui L, Yu FR, Yan Q. When big data meets software-defined networking: SDN for big data and big data for SDN. IEEE Network. 2016;30(1), 58–65. doi: 10.1109/MNET.2016.7389832
[11] Hu H, Wen Y, Chua TS, Li X. Toward Scalable Systems for Big Data Analytics: A Technology Tutorial. *IEEE Access*. 2014;2:652–687. Article 6842585. doi: 10.1109/ACCESS.2014.233245
[12] Feamster N, Rexford J, Zegura E. The Road to SDN. *ACM Networks*. 2013;11(12):20–40. Available: https://dl.acm.org/citation.cfm?doid=2559899.2560327
[13] "Netflix – Watch TV Shows Online, Watch Movies Online." [Online]. Available: https://www.netflix.com/
[14] Drago I, Mellia M, Munafo MM, Sperotto A, Sadre R, Pras A. Inside Dropbox. *Proceedings of the 2012 ACM Conference on Internet Measurement Conference* – IMC'12, New York, New York: ACM Press, Nov 2012, p. 481. [Online]. Available: https://dl.acm.org/citation.cfm?id=2398776.2398827
[15] "Amazon Web Services (AWS) – Cloud Computing Services." [Online]. Available: https://aws.amazon.com
[16] C. Metz. "The Epic Story of Dropbox's Exodus from the Amazon Cloud Empire." [Online]. Available: https://www.wired.com/2016/03/epic-story-dropboxs-exodus-amazon-cloud-empire/
[17] Sandvine - Intelligent Broadband Networks. "2016 Global Internet Phenomena Report - Latin America and North America." Tech. Rep., 2016. [Online]. Available: https://www.sandvine.com/hubfs/Sandvine_Redesign_2019/Downloads/Internet%20Phenomena/2016-global-internet-phenomena-report-latin-america-and-north-america.pdf
[18] Hadoop. "Hadoop Hive." 2014. [Online]. Available: https://hive.apache.org
[19] Hadoop. "Apache Hadoop." [Online]. Available: https://hadoop.apache.org

[20] Alqarni, MA. "Benefits of SDN for Big data applications," *2017 14th International Conference on Smart Cities: Improving Quality of Life Using ICT & IoT (HONET-ICT)*, Irbid/Amman, Jordan, 2017, pp. 74–77. doi: 10.1109/HONET.2017.8102206.
[21] Cisco. Cisco Visual Networking Index: Forecast and Methodology, 2015-2020 White Paper. Cisco Systems, Inc., San Jose, USA; 2016.
[22] Jarschel M, Wamser F, Hohn T, Zinner T, Tran-Gia P. SDN-based Application: Aware Networking on the Example of YouTube Video Streaming. *Proceedings of the 2nd European Workshop on Software Defined Networks (EWSDN 2013)*, Berlin, Germany, 2013. pp. 87–92.
[23] Dobrijevic O, Santl M, Matijasevic M. Ant colony optimization for QoEcentric flow routing in software-defined networks. *2015 11th International Conference on Network and Service Management (CNSM)*, IEEE, 2015. pp. 274–278.
[24] Zhu J, Vannithamby R, Rodbro C, Chen M, Andersen SV. Improving QoE for Skype video call in mobile broadband network. *Global Communications Conference (GLOBECOM)*, IEEE, 2012, pp. 1938–1943.
[25] Staehle B, Hirth M, Pries R, Wamser F, Staehle D. YoMo: A YouTube application comfort monitoring tool. *New Dimensions in the Assessment and Support of Quality of Experience for Multimedia Applications*, Tampere, Finland, 2010, pp. 1–3.
[26] Petrangeli S, Wauters T, Huysegems R, Bostoen T, De Turck F. Network-based dynamic prioritization of http adaptive streams to avoid video freezes. *Integrated Network Management (IM), 2015 IFIP/IEEE International Symposium on. IEEE*, 2015, pp. 1242–1248.
[27] Houdaille R, Gouache S. Shaping HTTP adaptive streams for a better user experience. *Proceedings of the 3rd Multimedia Systems Conference*, ACM, 2012, pp. 1–9.
[28] Bentaleb A, Begen AC, Zimmermann R. SDNDASH: Improving QoE of HTTP adaptive streaming using software defined networking. *Proceedings of the 2016 ACM on Multimedia Conference*, ACM, 2016, pp. 1296–1305.
[29] Nam H, Kim KH, Kim JY, Schulzrinne H. Towards QoE-aware video streaming using SDN. In: *Global Communications Conference (GLOBECOM), 2014 IEEE*, IEEE, 2014, pp. 1317–1322.
[30] Cofano G, De Cicco L, Zinner T, Nguyen-Ngoc A, Tran-Gia P, Mascolo S. Design and experimental evaluation of network-assisted strategies for HTTP adaptive streaming. *Proceedings of the 7th International Conference on Multimedia Systems*, ACM, 2016, p. 3.
[31] Joseph V, de Veciana G. NOVA: QoE-driven optimization of DASH-based video delivery in networks. *INFOCOM, 2014 Proceedings IEEE*, IEEE, 2014, pp. 82–90.
[32] Bahga A, Madisetti V. "Big Data Science & Analytics: A Hands-On Approach." 2016. Available: https://www.worldcat.org/title/ big-data-science-analytics-a-hands-on-approach/ oclc/953867348&referer=brief_results
[33] Cui L, Yu FR, Yan Q. When big data meets software-defined networking: SDN for big data and big data for SDN. *IEEE Network*. 2016;30(1):58–65.
[34] Mitchell T. *Machine Learning*. McGraw Hill; 1997.

[35] Hastie T, Tibshirani R, Friedman J. *The Elements of Statistical Learning.* Springer; 2001.
[36] Murphy KP. *Machine Learning: A Probabilistic Perspective.* MIT Press; 2013. Available: https://www.amazon.com/Machine-Learning-Probabilistic-Perspective-Computation/dp/0262018020/ref=sr_1_2?ie=UTF 8&qid=1336857747&sr= 8-2
[37] Han J, Kamber M. *Data Mining: Concepts and Techniques*, 2nd ed. Burlington, MA: Morgan Kaufmann; 2006. Available: https://www.elsevier.com/books/data-mining/han/978-0-12-381479-1
[38] Witten IH, Frank E, Hall MA, Pal CJ. *Data Mining: Practical Machine Learning Tools and Techniques.* Morgan Kaufmann; 2017. Available: https://www.worldcat.org/title/data-mining-practical-machine-learning-tools-and-techniques/oclc/972378340&referer=brief_results
[39] Maimon O, Rokach L. *Data Mining and Knowledge Discovery Handbook.* Vol. 14. Boston, MA: Springer; 2010. doi: 10.1007/978-0-387-09823-4.
[40] Berkhin P. A Survey of Clustering Data Mining Techniques. *Grouping Multidimensional Data*, pp. 25–71. Berlin/Heidelberg: Springer-Verlag; 2002. doi: 10.1007/3-540-28349-8_2
[41] Gama J. *Knowledge Discovery from Data Streams.* CRC Press; 2010. Available: https://www.worldcat.org/title/knowledge-discovery-from-data-streams/oclc/698165441&referer=brief_results
[42] Guo X, Singh S, Lee H, Lewis RL, Wang X. Deep Learning for Real-Time Atari Game Play Using Offline Monte-Carlo Tree Search Planning. In: Ghahramani Z, Welling M, Cortes C, Lawrence ND, Weinberger KQ, editors. *Advances in Neural Information Processing Systems 27.* Curran Associates, Inc.; 2014. pp. 3338–3346. Available: https://papers.nips.cc/paper/5421-deep-learning-for-real-time-atari-game-play-using-offline-monte-carlo-tree-search-planning.pdf
[43] Dean J, Ghemawat S. MapReduce: Simplified data processing on large clusters. *Communications of the ACM.* 2008;51(1):107–113.
[44] ITU-T Recommendation G.114. International Telecommunication Union; 1993.
[45] Orsolic I, Pevec D, Suznjevic M, Skorin-Kapov L. YouTube QoE Estimation Based on the Analysis of Encrypted Network Traffic Using Machine Learning. *Globecom Workshops (GC Wkshps), 2016 IEEE*, IEEE, 2016, pp. 1–6.
[46] Apache Software Foundation. “Apache Spark: Lightning-Fast Cluster Computing.” Available: https://spark.apache.org/
[47] Weldon MK. *The Future X Network: A Bell Labs Perspective.* CRC Press; 2016.

Chapter 9

# Usage of AI and machine learning techniques for solving potential problems in networks using SDN

*Ankush Mehra, Sumit Badotra, and Gurpreet Singh*

## 9.1 INTRODUCTION

Artificial Intelligence (AI) and machine learning (ML) approaches have been widely employed to solve various challenges, including routing, network classification, flow clustering, detection of intrusions, distribution of load, fault detection, Quality of Service and Quality of Experience optimization, admission control, and resource allocation. However, in the era of SDN, the role of AI has grown dramatically as a result of the enormous efforts made by the industry and research communities [1]. Recent studies have revealed a strong trend in the scientific community to utilize AI methods in SDNs.

### 9.1.1 Software-defined network

Network control logic is separated from the hardware that carries it out, like routers, to regulate data flow within the underlying network through software-defined networking (SDN). Infrastructure management, which may be divided into multiple organizations or specialized for a single business, is made easier with this method. SDN allows policy-based management through controllers that sit on top of the network hardware, whether it's in the cloud or on premises. Considering that the forwarding and network control planes are technically separate from the data plane or underlying infrastructure, the organization can implement network control directly [1]. This environment is quite different from a typical data center arrangement. In usual circumstances, a router or switch in the cloud or physically in the data center is aware only of network devices close to it. SDN provides intelligence that may be broadly distributed and centralized, allowing it to observe and handle everything [1].

### 9.1.2 Growing complexities and challenges in modern network environments

In the present era, contemporary network environments grapple with an escalating array of complexities and challenges, necessitating sophisticated

 DOI: 10.1201/9781003504559-9

solutions. The increasing adoption of connected gadgets, the appearance of contemporary technologies including the Internet of Things, and the developing reliance on cloud-based totally definitely offerings have all extended the breadth and problem of networks. The extended diploma of element not handiest exposes networks to new vulnerabilities however also demanding situations them with increasingly more complicated and varied cyber assaults [2]. Furthermore, the developing want for speedy connectivity and seamless integration for the duration of several systems places greater pressure on network infrastructure. Network directors face the daunting burden of assuring the network's security and resilience at the same time as also ensuring its adaptability to fast technological advances. In the exploration of reaping the benefits of virtual transformation, efficaciously navigating via those complexities calls for a complete method that integrates strong cyber safety features, green community management, and strategic making plans to successfully cope with the multifaceted traumatic conditions inherent in modern community environments [3] as shown in the Table 9.1.

*Table 9.1* Growing complexities and challenges in modern network environments

| *Challenges* | *Description* |
|---|---|
| Increasing network traffic | The total volume of traffic is still growing rapidly as more devices and apps connect to networks. This reduces the network's performance and capacity. |
| Security threats | Cybersecurity risks such as ransomware, malware, Distributed Denial of Service assaults, and others are always present on networks. Network security is a never-ending problem. |
| Supporting new technologies | The emergence of technologies such as IoT, 5G, WiFi 6, SD-WAN, etc. presents both opportunities and challenges in terms of support. |
| Management and monitoring | With increasing traffic, scale, and dynamic changes, networks are more difficult to maintain. While helpful, automation and artificial intelligence are difficult to use. |
| Legacy systems | There are still many outdated switches, routers, protocols, etc. in use. There is risk and expense associated with replacing or upgrading them. |
| Lack of skilled staff | Insufficient qualified network engineers and IT personnel who are up to date on current technology. |
| Compliance requirements | Complying with stringent requirements regarding data security, privacy, and other issues is necessary yet difficult. |
| Business alignment | Networks are seen as a cost center, not a strategic asset. Lack of alignment with business needs. |

### 9.1.3 The growing need for machine learning and AI in network management

To effectively address increasing complexity, legacy network management techniques are no longer adequate [4]. The utilization of AI and ML is imperative to facilitate automated and intelligent network administration.

Traditional network monitoring and troubleshooting techniques are labor-intensive and lead to significant delays and inefficiencies. The enormous amounts of data flow and events produced across diverse networks in real time are too big for human operators to handle alone [5]. Problems are discovered and fixed reactively, after the effects have already been felt, in the absence of proactive self-healing and predictive skills. This is where network operations can be completely transformed by AI and ML.

AI and ML models can instantly identify regular traffic patterns and discover anomalies by applying algorithms to big datasets. Predictive analytics enables enhanced performance and security along with quick incident response. Networks can become self-adjusting systems that constantly learn and modify policies to suit business goals, as opposed to merely being reactive break-fix systems [6]. The integration of AI automation and human experience in augmented intelligence solutions frees network engineers to concentrate on strategic, high-value tasks rather than tedious manual labor.

AI-powered solutions for network administration, assurance, security, and other applications are now available from top networking providers. As the technology advances, the number of use cases will increase. ML is being used by cloud-based management solutions to provide automated insights and actions [7]. Intelligent analytics are being provided to SDN controllers for quick failure diagnosis and dynamic traffic steering. In addition to transforming wireless networks, AI is also clearing the path for more effective self-organizing 5G systems by enhancing transmission quality.

AI holds enormous potential for improving network operations. To optimize value, however, a few significant obstacles must be addressed, including poor data quality, algorithmic biases, and a shortage of qualified personnel [8]. The exponential development in network complexity and capacity demands can only be met with the help of AI and ML, together with appropriate governance and human–machine collaboration. Future networks that are self-optimized, robust, and able to withstand the rapid advancement of digital technology will be built on intelligent automation [9].

## 9.2 SDN (SOFTWARE-DEFINED NETWORKING) ARCHITECTURE

The network control plane and data plane are divided by SDN [10]. This makes it possible to separate infrastructure from applications and network services and to program the network.

The SDN architecture consists of three layers (Figure 9.1):

1. **Application Layer:** Consists of SDN applications that use APIs to connect with the controller and share goals and specifications. Traffic engineering, load balancing, and security apps are a few examples [10].
2. **Control Layer:** The network's central nervous system. Using southbound APIs such as OpenFlow, the SDN controller manage flow to establish the forwarding choices that network devices receive. Converts application needs into commands for the network [10].
3. **Infrastructure Layer:** Comprises the physical components of a network, such as switches, routers, and gateways. They may be programmed via SDN controllers utilizing southbound APIs, and they offer packet forwarding data-plane functionality [10].

## 9.2.1 Separation of control and data planes in SDN architecture

In a network, the control plane decides how to calculate routes and forward packets. It manages the network's cognitive functions and decision-making.

The network hardware that forwards traffic, such as switches and routers, is known as the data plane. It handles the processing and transmission of packets[11].

Conventional networks perform both data forwarding and control logic (routing protocols, signaling, etc.) in a distributed manner on similar devices. The network is inflexible and difficult to modify [11].

SDN divides these two layers:

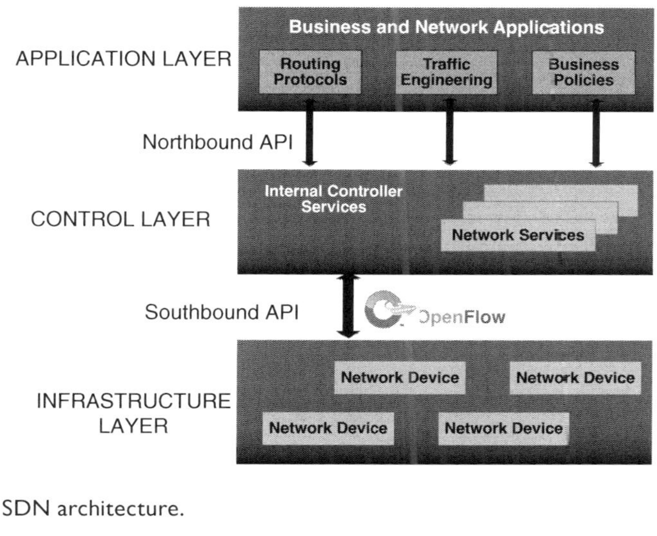


*Figure 9.1* SDN architecture.

- In software-based SDN controllers, the control plane is centralized.
- Hardware components like switches and routers still have data planes.

The control logic and the supporting infrastructure are separated by this division. This allows for flexible programming of the control plane without requiring direct contact with particular hardware components. The two layers of SDN are shown in Figure 9.2. SDN controllers can control data plane devices forwarding policies and have a worldwide perspective of the network. Networks are now dynamic and centrally managed as a result [11,12]. Programming the SDN controller allows for the deployment of new features without having to update every distributed device. Vendor-agnostic implementations are made possible by support for open APIs such as OpenFlow [12].

The strategic benefits of SDN, as shown in Figure 9.3—namely network programmability, automation, and flexibility—primarily stem from the separation of the control and data planes [12]. It does away with the need for vendor-specific hardware updates.

### 9.2.2 The role of the SDN controller in network management

The primary software application that automates and programmatically regulates network activity is called the SDN controller. It acts as the SDN architecture's intellect [13].

The SDN controller's principal duties consist of:

- Preserving a global perspective of every linked device and the whole network topology. Centralized control is made possible by this.
- Offering a layer of abstraction between the underlying infrastructure and network services. It allows for flexibility.

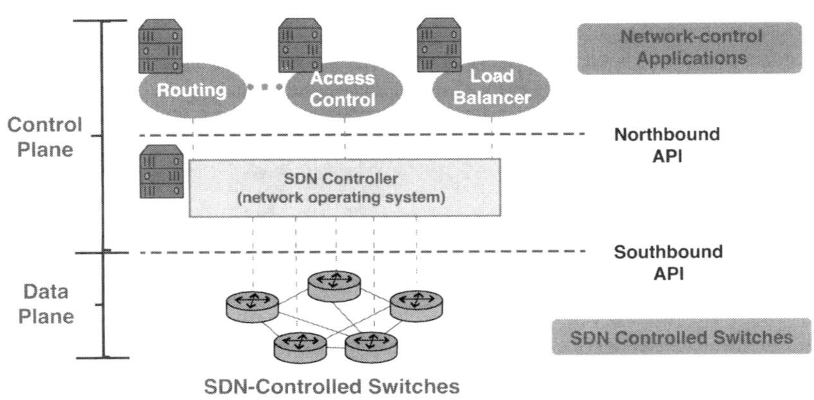


*Figure 9.2* SDN divided into two layers.

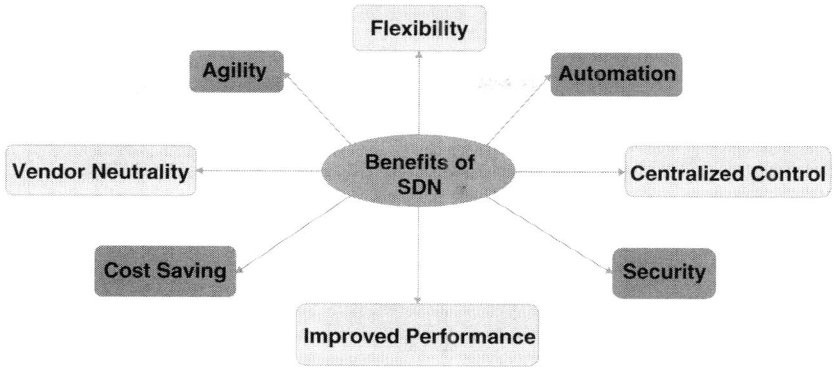


*Figure 9.3* Benefits of SDN.

- Applying business logic and high-level policies to traffic flows such that they are determined by intent rather than by device configurations.
- Putting forwarding logic and rules into data-plane hardware, such as switches and routers.
- Transforming application needs into network commands to enable communication between the application layer and network devices.

## 9.3 THE PRESENT STATE OF NETWORK ISSUES

Today's network infrastructures deal with a wide range of challenging problems brought on by the quick speed of technological advancement. New developments like cloud, Internet of Things, mobility, and 5G are difficult to support on legacy network infrastructure with rigid architectures [14]. Massive traffic spikes are straining network capacity due to the increasing diversity of devices, users, and applications. The increasing sophistication of security threats necessitates constant observation. While the demands for uptime and performance rise, networks become more difficult to manage due to a lack of automation and intelligence. These difficulties are made worse by a lack of competent talent [14]. Networks are not flexible enough to adapt to changing business needs due to manual settings, device-centric management, complicated protocols, and vendor dependencies. More intelligent, autonomous, and adaptable networks are desperately needed so they may continue to function safely and dependably at scale as needs and technology change. Moving forward, fundamental adjustments to network management, operations, and architecture will be necessary [15].

### 9.3.1 Typical problems in traditional network configurations

- **Manual Configuration:** Device-by-device configuration and setup are manual processes for networks. This is laborious, prone to mistakes, and not very scalable [15].
- **Vendor Lock-in:** Multi-vendor setups are difficult to integrate because of vendor-specific equipment and protocols, which lead to dependencies [16].
- **Lack of Agility:** Reconfiguring individual devices is necessary when implementing new services or updates, and this process is time-consuming [16].
- **No Centralized View:** There is limited visibility into the overall functioning of the network in the absence of centralized control and monitoring [16].
- **Complex Troubleshooting:** Identifying and resolving issues among dispersed devices with diverse configurations is challenging [16].
- **Security Issue:** Across traditional networks, implementing uniform and flexible security standards is difficult.
- **Rigid Architectures:** Hierarchical networks and other outdated systems are too inflexible to effectively handle changing needs [16].
- **Limited Automation:** Organization and administration are difficult due to the limited built-in automation features.
- **Scaling Constraints:** At bigger levels, traditional networks don't scale well in terms of durability, performance, or capacity [16].
- **Lack of Programmability:** Flexibility is hampered by the inability to modify the network dynamically through software programmability.
- **Siloed Operations:** Network domains, IT systems, and applications are not integrated [16].

## 9.4 AI (ARTIFICIAL INTELLIGENCE) AND ML (MACHINE LEARNING)

### 9.4.1 Artificial intelligence and machine learning

AI systems are software programs developed to do actions that would ordinarily need human intelligence. AI systems are designed to simulate cognitive functions such as perception, learning, reasoning, problem-solving, and prediction [17].

ML, a subtype of AI, enables algorithms to automatically learn from data and improve their performance at specific tasks over time without requiring explicit programming [17]. ML aims to construct programs that can learn from experience and adapt as they encounter new data.

Among AI/ML's primary functionalities are [18]:

- **Pattern recognition** - finding patterns and irregularities in big datasets.
- **Predictive modeling** - forecasting and making predictions based on data.
- **Classification** - categorizing inputs into groups based on identified patterns.
- **Natural language processing** - recognizing the subtleties of human language.
- **Computer vision** - correctly recognizing objects in pictures or videos.

AI/ML is being applied extensively in a variety of fields, including stock trading, fraud detection, digital assistants, self-driving automobiles, and medical diagnostics [18]. Network operations, security, traffic optimization, and other areas are becoming more intelligent and automated thanks to AI/ML. Data volumes will increase, and managing complexity and gaining insights will depend more and more on AI and ML.

### 9.4.2 Key differences and similarities of AI and ML

Creating intelligent computers that can think like humans is the shared objective of two well-known computer science disciplines: AI and ML. Instead of using explicit programming, these domains use statistical models and techniques to let machines learn from data [19]. But AI is a more general term for robots that exhibit intelligence. It includes a wide range of methods, including planning, prediction, reasoning, problem-solving, and ML [19]. In contrast, ML focuses on pattern identification and predictive analytics by creating algorithms that are capable of learning and improving over time.

While ML focuses primarily on building statistical models based on large datasets to perform specific tasks like recognition, classification, and prediction without explicit instructions, AI uses symbolic logic, hard-coded rules, knowledge bases, and ML to mimic a wide range of human cognitive skills [20]. Some AI techniques, in contrast to ML, can rely more on expert systems and rules that have been handpicked by human engineers rather than vast training datasets. While some AI systems are built on intelligible symbolic logic and reasoning, ML models typically function as opaque black boxes. While they both investigate the automation of complex intellect, AI strives for greater general intelligence, whereas ML excels at specialized tasks [20].

Here are some differences and similarities of AI and ML as shown in Tables 9.2 and 9.3.

*Table 9.2* Differences between artificial intelligence and machine learning

| *Differences* | *Descriptions* |
|---|---|
| Scope | Machine Learning (ML) is a subset of artificial intelligence (AI), which is a larger field. Statistical learning algorithms are the particular emphasis of ML. |
| Approach | AI can include knowledge representation, expert systems, rule-based programming, symbolic logic, and more. Data-driven statistical methods are the mainstay of machine learning. |
| Recognizing | ML algorithms automatically pick up new skills from data. Human-designed hard-coded rules can be incorporated into AI systems. |
| Functionality | Predictive analytics is ML's main emphasis. AI seeks to automate more cognitive functions that humans possess. |
| Data dependency | Large volumes of training data are needed for ML algorithms. Certain AI systems depend more on curated knowledge sources and guidelines. |
| Interpretability | ML models are frequently opaque. There are AI systems that rely on intelligible symbolic thinking. |
| Generalization | ML excels at a few specialized tasks. The goal of AI is more general intelligence. |
| Logic | Complex patterns are recognized using ML. Additional methods including knowledge representation, planning, reasoning, and problem solving are incorporated into AI. |

*Table 9.3* Similarities between artificial intelligence and machine learning

| *Similarities* | *Descriptions* |
|---|---|
| Data-driven approaches | They mostly rely on statistical models, pattern identification from big datasets, and data representation. |
| Intent to imitate human intellect | Developing intelligent software and systems with abilities like sensing, creativity, problem-solving, reasoning, and learning is the ultimate goal of both areas. |
| Automating difficult procedures | Automating difficult tasks that have traditionally demanded human intelligence is one of the main goals. |
| Adaptability | Their goal is to develop systems that are capable of learning from experience and adapting over time. |
| Multidisciplinary | Computer science, linguistics, mathematics, psychology, neuroscience, philosophy, and other disciplines are all integrated in these fields of study. |
| Prediction | Both entail creating methods and algorithms for data-driven forecasting and well-informed decision-making. |
| Black-box modeling | The produced models may be opaque "black boxes" with confusing internal mechanisms. |
| State-of-the-art technology | Innovative, cutting-edge technologies like artificial intelligence and machine learning have a wide range of practical uses. |

### 9.4.3 Applications of artificial intelligence and machine learning in network security

- **Applications of AI:** AI in information security studies usage trends to detect potentially risky people or acts and predict cyberattacks [21]. AI-powered automatic monitoring ensures system security around the clock and allows organizations to take action before damage occurs. Some of the most important applications of AI in network security include:
  1. Malware Detection
  2. Intrusion Detection System
  3. Spam Filtering
  4. DDoS Attack Protection
  5. Network Behavior Analysis
  6. Automated Threat Analysis
  7. Securing IoT Devices
  8. User and Entity Behavior Analytics

- **Applications of ML:** The most important aspect of every business is cybersecurity. It aids in ensuring the security and safety of their data. AI and ML are in high demand, and this is changing cybersecurity as well. ML is extremely useful in cybersecurity since it can be used to detect threats, combat cybercrime, and improve antivirus software by leveraging AI capabilities [21]. So, let's analyze some ML applications in cybersecurity:
  1. Anomaly Detection
  2. Malware Classification
  3. Spam Filtering
  4. DDoS Attack Mitigation
  5. Network Intrusion Detection
  6. Fraud Detection
  7. Automated Threat Intelligence
  8. Encryption Technique

## 9.5 A COMPARISON OF MACHINE LEARNING TECHNIQUES FOR DETECTING DDOS ATTACKS

This comparative study will investigate numerous ML techniques and their advantages, disadvantages, and suitability for SDN Distributed Denial of Service (DDoS), as shown in Table 9.4 [22, 23].

The table outlines the benefits, dangers, and applicability of various device learning techniques for detecting DDoS attacks in SDN structures [22,23]. Techniques inclusive of artificial neural networks, convolutional

*Table 9.4* Comparative analysis of machine learning techniques for DDoS attacks

| *Techniques* | *Advantages* | *Disadvantages* | *Suitability for SDN DDoS* |
|---|---|---|---|
| Support Vector Machines | Effective for high-dimensional data, resilient against noise | Sensitive to parameter adjustment; computationally expensive | Moderate and suitable for static attacks |
| Decision Trees & Random Forests | Accessible, handling non-linear data, and resistant to noises | Can overfit and bias for high cardinality features | Moderate and suitable for static attacks |
| Clustering (K-Means, DBSCAN) | Unsupervised, can detect new assaults | Difficult to interpret clusters, vulnerable to anomalies. | Moderate and suitable for static attacks |
| Convolutional Neural Networks | Effective for spatial features and automatic feature extraction | Require huge training data, computationally expensive | High, appropriate for dynamic attacks |
| Ensemble Methods (Bagging, Boosting, Stacking) | enhanced accuracy and resilience | Increased complexity can be computationally expensive | High, appropriate for dynamic attacks |
| Hybrid Approaches (Combining techniques) | Utilizes the strengths of several methodologies | Increased complexity may require careful tuning | High, appropriate for dynamic attacks |
| Artificial Neural Networks | Effective and adaptive for complex patterns | Large amounts of training data are required, which can be computationally costly | High, appropriate for dynamic attacks |

neural networks, SVMs, ensemble methods, and hybrid approaches with SDN-particular capabilities are more effective at detecting dynamic and complex DDoS assaults in SDN structures, but may additionally require larger training datasets and processing resources.

## 9.6 SDN ENVIRONMENT DDOS DETECTION AND DEFENSE

A trigger mechanism to detect potential attacks, a controller that uses a hybrid ML approach (K-Means and KNN) to identify suspicious flows, makes up the proposed DDoS detection and defensive system, and a specific mechanism to fend off DDoS attacks. Figure 9.4 illustrates the framework's high-level depiction.

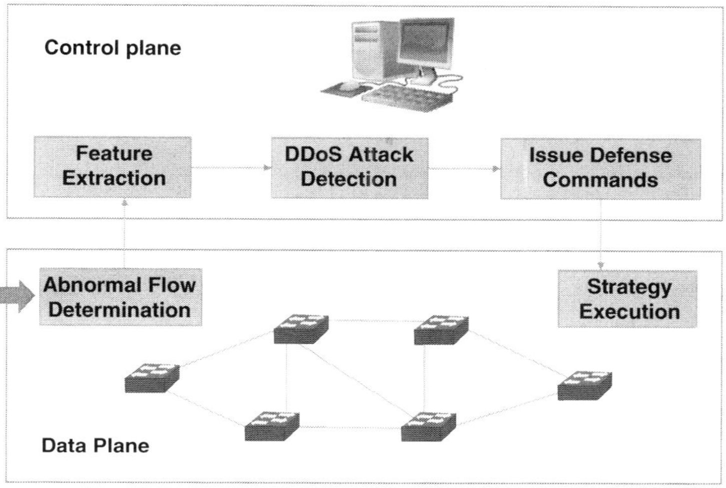


*Figure 9.4* How the SDN environment's trigger mechanism, cetection, and mitigation cooperate.

### 9.6.1 A mechanism for detecting DDoS attacks on a programmable data plane

Effective DDoS detection techniques that monitor network traffic at predetermined intervals have been proposed by researchers. However, this approach strains the controller's performance and results in delayed responses, overwhelming packet load for both controllers and switches. This burdens the network and increases pressure during subsequent DDoS attack defenses if response time isn't swift.

Shorter detection intervals in DDoS detection methods in SDN environments lead to frequent attacks, monopolizing controller resources and increasing communication overhead, affecting concurrent processes, and highlighting the shortcomings of periodic trigger-based methods [24].

The paper introduces a trigger mechanism for DDoS attack detection, overcoming traditional trigger-based methods' limitations and addressing workload and resource utilization concerns in switches in SDN, which primarily function as data-plane forwarding devices [24].

Optimizing idle resources on SDN switches can reduce the controller's detection workload without affecting switch performance. By deploying a detection trigger mechanism, this strategy enhances detection capabilities, optimizes network efficiency, and minimizes the controller burden [24].

The study aims to reduce controller resource consumption and communication overhead by putting in place a data-plane detecting trigger system, as switches typically detect DDoS attacks after they do so, thereby enhancing overall system performance [25].

Attackers employ packet_in message counting on switches to counter DDoS attacks. Flow entries and source and destination IP addresses of switches are altered as a result of the attacks, inundating networks with fictitious IP addresses.

Switches have trouble figuring out the best paths and ask the controller for forwarding paths. The controller processes packet_in messages as a result [25].

In response to packet_in messages, the controller makes changes to switches or sends out packet_out messages to instruct switches on how to operate properly. When the controller does not respond, switches keep sending packet_in messages to it, which results in an endless loop, as illustrated in Figure 9.5.

The volume of packet_in messages that switches forward dramatically increases during a denial-of-service assault that targets the controller. This surge in packet_in message rates serves as an indicator of potential DDoS attacks. Detecting an abnormal increase in packet_in messages signals the existence of suspicious flows within the switches, prompting the controller to activate the attack detection process. Figure 9.6 outlines the algorithmic approach proposed in this article for the detection trigger mechanism [24].

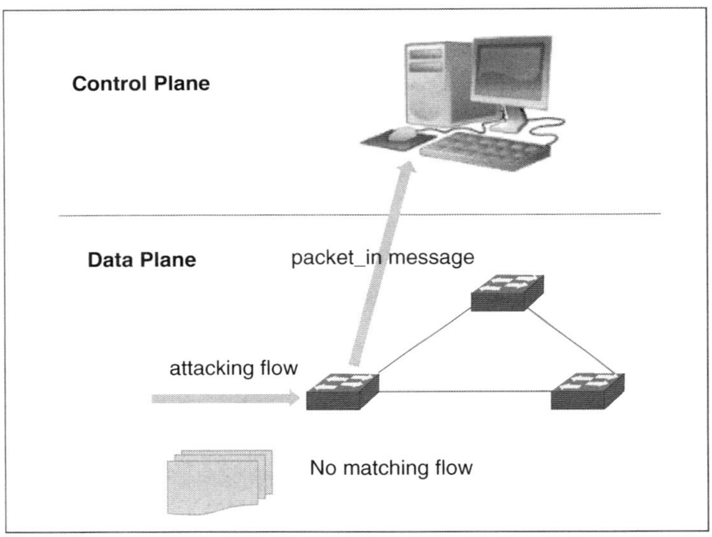


*Figure 9.5* Detailed illustration of a DDoS attacks.

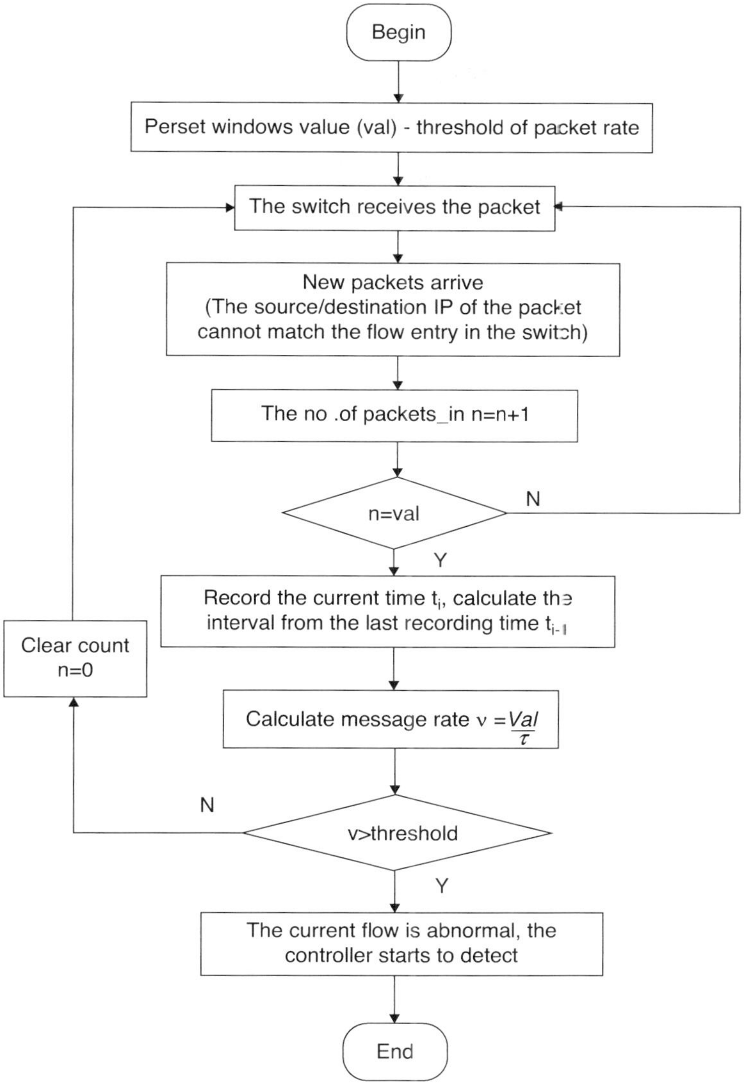


*Figure 9.6* Process of identifying the trigger mechanism.

## 9.6.2 Flow feature analysis

The features utilized for flow classification have a big influence on the algorithm's accuracy in classification-based DDoS detection techniques. To more precisely distinguish attack flows from regular ones, this section goes into great detail on how attackers initiate DDoS attacks on SDN controllers [24,25]. DDoS attacks disrupt network operations by crashing equipment or congesting links, impeding normal service delivery. They flood the network with large amounts of traffic, leading to higher average throughput through switches. The key distinctions between regular and DDoS attack flows are the typical length and byte rate of the stream, enabling the distinction between malicious and benign network traffic.

The asymmetry of traffic flow is a crucial feature in identifying DDoS attacks. Typically, bidirectional traffic results in a high percentage of symmetric traffic. However, during a DDoS attack, the number of individual flows increases significantly, decreasing the proportion of symmetric flows. This is similar to genuine traffic burst activity, making it essential to distinguish between malicious and normal traffic surges. Normal traffic patterns and DDoS-induced traffic fluctuations can be distinguished from one another by utilizing the asymmetric traffic's rate of change [22]. Attackers often transmit a small number of packets, which indicates a potential DDoS attack. Assessing the proportion of flows with minimal packets is a reliable indicator of network traffic's state and the likelihood of a DDoS attack. Understanding this proportion can guide the level of defense needed to counter varying intensities of DDoS attacks effectively [22–25].

We've selected these traffic features to spot DDoS attacks from the data we've seen earlier, which is shown below step by step:

1. The average number of bytes per unit time at $T_n$ is calculated as the average number of bytes of flows per unit time.

$$br = \frac{b_{Tn} - b_{T_{n-1}}}{T_n - T_{n-1}} \tag{9.1}$$

2. Average durations per flow: the average time each flow takes.

$$adf = \frac{\sum_{i=0}^{\text{flow\_nums}} dur_i}{\text{flow\_nums}_T} \tag{9.2}$$

3. Percentage of pair-flows: the ratio of symmetric flow to total flows per unit time.

$$ppf = \frac{\sum_{i=0}^{\text{flow nums}_T} N_{\text{pair\_flows}}}{\sum_{i=0}^{\text{flow\_nums}} N_{\text{pair\_flows}} + \sum_{i=0}^{\text{flow\_nums}} N_{\text{single\_flows}}} \tag{9.3}$$

4. Variation rate of single flows: adjust the asymmetric flow rate per unit time.

$$vrsf = \frac{\text{flow\_nums}_T - \sum_{i=0}^{flow-mums_T} N_{\text{single\_nows}}}{T_n - T_{n-1}} \tag{9.4}$$

5. The percentage of flows with a few packets is the proportion of all flows with a few packets.

$$pfsp = \frac{\sum F_i\left(N_p < V\right)}{\text{flow\_ nums}_{\,T}} \tag{9.5}$$

In-band telemetry is a technology in SDN that collects data about packet movement in a network. It involves programmable switches recording information like ID, queue delay, departing time, packet size, and a "five-tuple" consisting of the protocol type, port numbers, source, and destination IP addresses. This data is used to calculate metrics and detect unusual traffic. A controller then analyzes the flow by sending standard messages and applying formulas to determine specific characteristics, which are saved in a database for future reference [25].

### 9.6.3 DDoS detection algorithm

By merging the K-nearest neighbor (KNN) and K-Means algorithms, a novel method for DDoS detection seeks to achieve a trade-off between speed and precision. The KNN method establishes the proximity of traffic to these groups, whereas the KNN-based traffic detector group comparable data and the K-Means-based training data processor [26]. If the nearest groups classify traffic as normal or an attack, a quick determination can be made. If not, distances between traffic and training data are calculated to find the closest matches as shown below.

Pseudo code for K-Means training data processing.

**Input:** Training Dataset

$$D_{\text{train}} = \left(X_1^{\text{label}}, X_2^{\text{label}}, \ldots, X_N^{\text{label}}, X_{N+1}^{\text{label}}, X_{N+2}^{\text{label}}, \ldots, X_{N+M}^{\text{label}}\right)$$

The number of normal data is N, while the quantity of abnormal data is M.

Each dataset has five-dimensional features.

$$X_i^{\text{label}} = \left(x_{i1}, x_{i2}, x_{i3}, x_{i4}, x_{i5}\right)$$

**Output:** Cluster class C= (C1, C2, ... Ck), centroid set Cen, radius set

1. **for** i = **1:** $N+M$ **do**
2. **for** j = 1:5 **do**
3. Normalized $x'_{ij} = \dfrac{x_{ij} - min\{x_j\}}{max\{x_j\} - min\{x_j\}}$
4. **end for**
5. **end for**

6. Randomly select k initial centers $X_{c1}^{\text{label}}, X_{c2}^{\text{label}}, \ldots, X_{ck}^{\text{label}}$
7. **repeat**
8. **for i = 1: *N+M-K* do**
9. **for** s = 1: k **do**
10. Calculate the distance between the sample $X_i^{\text{label}}$ of each cluster

$$D_{is}\left(X_i^{\text{label}}, X_{cs}^{\text{label}}\right) = \sqrt{\left(X_i^{\text{label}} - X_{cs}^{\text{label}}\right)^2} = \sqrt{\sum_{j=1}^{5}\left(x_{ij} - x_{cj}\right)^2}$$

11. According to the distance $D_{is}$, the sample $X_i^{\text{label}}$ is classified into the nearest cluster
12. **end for**
13. **end for**
14. **for** s=1: k **do**
15. Calculate new cluster center vector $(X_{cs}^{label})$`
16. **if** $\left(X_{cS}^{\text{label}}\right)' = X_{cS}^{\text{label}}$ **then**
17. Keep the current mean vector unchanged
18. **else**
19. update the current cluster center vector $X_{CS}^{\text{label}}$ to $\left(X_{CS}^{\text{label}}\right)'$
20. **end if**
21. **end for**
22. **until** the current cluster center vector is not updated
23. **for** s=1: k **do**
24. calculate the radius of each cluster $r_S = \dfrac{\sum_{i=1}^{N_S} D_{iS}}{N_S}$
25. **end for**

The KNN algorithm, which is used to detect DDoS attacks, is streamlined by using the training data processor's results, reducing the time spent on comparisons. This method maintains accuracy and efficiency by using the K-Means algorithm during training. K-Means group similar traffic data, making it easier for the algorithm to match features quickly. The detection process starts by normalizing the observed data, the nearest clusters to the detected flow are identified by measuring their distance from one another. If these clusters label the flow as normal or abnormal, the method is quick. By considering the ratio of normal to abnormal points, we reduce computation cost, enhance algorithm efficiency, and reduce the effect of anomalous points on the outcome [27].

### 9.6.4 Analysis using simulated data

We concentrate on recall, accuracy, and mistake rate as we thoroughly assess the effectiveness and usability of the DDoS attack detection system in a network simulation with SDN. To evaluate the performance of our strategy, we compared two benchmarking methodologies: a self-organizing map (SOM)-based ML detection algorithm and JESS [4], which employs joint entropy [8]. DDoS detection systems in SDN are typically classified into two types: those based on entropy and those that use ML. In JESS [4] first, we use a combined entropy technique, which takes into account both the IP address's entropy and the TCP attributes when paired with the IP address. Nonetheless, it struggles to distinguish DDoS attack flows from abrupt bursts because entropy values only indicate data randomness. As a result, its detection accuracy is frequently lower than that of ML-based systems.

In contrast, our second comparison, Distributed-SOM [8], is a ML-powered DDoS detection solution that uses distributed SOMs to tackle DDoS flooding. This strategy significantly improves detection accuracy and speed while putting no burden on the network infrastructure (Table 9.5).

In comparison to the ML-based method, we found that the entropy-dependent identification method had much lower accuracy and recall rates (93.79% and 92.80%, respectively). This gap shows that ML-based solutions are more effective at identifying DDoS attack flows. The entropy-driven technique has a much higher misjudgment speed (6.95%) than the Distributed-SOM method (1.75%) or our proposed detection method (0.97%). Entropy-based detection is based purely on data unpredictability, and it frequently misidentifies oscillations in network services as DDoS attacks. ML-based techniques, on the other hand, achieve higher levels of accuracy due to their more robust detection algorithms. Our suggested detection method outperforms previous ML techniques because of its detection trigger mechanism, which pre-filters network traffic, allowing controllers to concentrate on examining questionable data.

## 9.7 CONCLUSION

Although SDN offers numerous advantages, it may be vulnerable to DDoS attacks, a common threat to network security. Because SDN is centralized, its controller could be exposed to DDoS attacks. SDN causes network changes by rethinking network management and connection. Although SDN has many benefits, it also introduces new security challenges that academics and industry must address. DDoS assaults can harm both the SDN control and data planes. DDoS assaults are more difficult to detect today

*Table 9.5* Test findings using various techniques

| *Detective method* | *Accuracy* | *Recall* | *False positive* |
|---|---|---|---|
| Distributed - SOM | 98.47 | 97.78 | 1.75 |
| Entropy method | 93.79 | 92.79 | 6.95 |
| Proposed method | 98.85 | 98.47 | 0.97 |

than in the past due to the complexities of traffic flow. The experiment demonstrates the effectiveness of ML in detecting DDoS attacks inside network data. The new study reveals how ML algorithms can detect and mitigate DDoS attacks, paving the path for more effective network security solutions. Experimentation demonstrates the utility of the proposed detection methodologies, highlighting the trigger mechanism's efficiency in identifying aberrant flows while conserving controller resources.

## REFERENCES


1. Xie, J., Guo, D., Hu, Z., Qu, T., Lv, P. Control plane of software defined networks: A survey. *Computer Communications*, 2015, 67, 1–10.
2. Suciu, G., Vulpe, A. An overview study of software defined networking. In *Proceedings of the IE International Conference*, Bucharest, Romania, 30 April–3 May 2015.
3. Kreutz, D., Ramos, F., Verissimo, P. Towards secure and dependable software-defined networks. In *Proceedings of the Second ACM SIGCOMM Workshop on Hot Topics in Software Defined Networking*, New York, NY, USA, 16 August 2013.
4. Sundas, A., Badotra, S. Comprehensive study of machine learning-based systems for early warning of clinical deterioration. *International Journal of Performability Engineering*, 2022, 18(12), 893.
5. Shuaib, M., Badotra, S., Khalid, M. I., Algarni, A. D., Ullah, S. S., Bourouis, S., ... & Gundaboina, L. A novel optimization for GPU mining using overclocking and undervolting. *Sustainability*, 2022, 14(14), 8708.
6. Badotra, S., Panda, S. N. SNORT based early DDoS detection system using Opendaylight and open networking operating system in software defined networking. *Cluster Computing*, 2021, 24, 501–513.
7. Gundaboina, L., Badotra, S., Tanwar, S. Reducing resource and energy consumption in cryptocurrency mining by using both proof-of-stake algorithm and renewable energy. In *2022 International Mobile and Embedded Technology Conference (MECON)*, pp. 605–610, IEEE, Noida, Uttar Pradesh, 2022.
8. Badotra, S., Panda, S. N. Evaluation and comparison of OpenDayLight and open networking operating system in software-defined networking. *Cluster Computing*, 2020, 23, 1281–1291.
9. Mostafaei, H., Menth, M., Obaidat, M. S. A learning automaton- based controller placement algorithm for software-defined networks. In *Proceedings of the 2018 IEEE Global Communications Conference (GLOBECOM)*, Abu Dhabi, UAE, 9–13 December 2018.
10. Badotra, S., Panda, S. N. SNORT based early DDoS detection system using Opendaylight and open networking operating system in software defined networking. *Cluster Computing*, 2021, 24(1), 501–513.
11. Mohan, M., Tamizhazhagan, V., Balaji, S. A perspicacious multi-level defense system against DDoS attacks in cloud using information metric & game theoretical approach. *Journal of Network and Systems Management*, 2023, 31(4), 1–28.

12. Alshahrani, M. M. A secure and intelligent software-defined networking framework for future smart cities to prevent DDoS attack. *Applied Sciences*, 2023, 13(17), 9822.
13. Gundaboina, L., Badotra, S., Tanwar, S. Reducing resource and energy consumption in cryptocurrency mining by using both proof-of-stake algorithm and renewable energy. In *2022 International Mobile and Embedded Technology Conference (MECON)*, pp. 605–610, IEEE, 10 March 2022.
14. Paidipati, K. K., Kurangi, C., Uthayakumar, J., Padmanayaki, S., Pradeepa, D., Nithinsha, S. Ensemble of deep reinforcement learning with optimization model for DDoS attack detection and classification in cloud based software defined networks. *Multimedia Tools and Applications*, 2023, 83, 1–19.
15. Aljebreen, M., Mengash, H. A., Arasi, M. A., Aljameel, S. S., Salama, A. S., Hamza, M. A. Enhancing DDoS attack detection using snake optimizer with ensemble learning on Internet of Things environment. *IEEE Access*, 2023, 11, 104745–104753.
16. Gupta, N., Maashi, M. S., Tanwar, S., Badotra, S., Aljebreen, M., Bharany, S. A comparative study of software defined networking controllers using mininet. *Electronics*, 2022, 11(17), 2715.
17. Ramalakshmi, R., Kavitha, D. DDoS attack mitigation using distributed SDN multi controllers for Fog based IoT systems. *International Journal of Intelligent Systems and Applications in Engineering*, 2024, 12(4s), 57–69.
18. Kumar, A., Dutta, S., Pranav, P. FQBDDA: Fuzzy Q-learning based DDoS attack detection algorithm for cloud computing environment. *International Journal of Information Technology*, 2023, 83, 1–10.
19. Badotra, S., Nagpal, D., Panda, S. N., Tanwar, S., Bajaj, S. IoT-enabled healthcare network with SDN. In *2020 8th International Conference on Reliability, Infocom Technologies and Optimization (Trends and Future Directions) (ICRITO)*, pp. 38–42, IEEE, 4 Jun 2020.
20. Chakraborty, S., Banerjee, S. Proposed approach to detect distributed denial of service attacks in software defined network using machine learning algorithms. *International Journal of Engineering and Technology*, 2018, 7, 472–476.
21. Shuaib, M., Badotra, S., Khalid, M. I., Algarni, A. D., Ullah, S. S., Bourouis, S., Iqbal, J., Bharany, S., Gundaboina, L. A novel optimization for GPU mining using overclocking and undervolting. *Sustainability*, 2022, 14(14), 8708.
22. Bharany, S., Badotra, S., Sharma, S., Rani, S., Alazab, M., Jhaveri, R. H., Gadekallu, T. R. Energy efficient fault tolerance techniques in green cloud computing: A systematic survey and taxonomy. *Sustainable Energy Technologies and Assessments*, 2022, 53, 102613.
23. Hu, D., Hong, P., Chen, Y. FADM: DDoS flooding attack detection and mitigation system in software-defined networking. In *GLOBECOM 2017: 2017 IEEE Global Communications Conference*, pp. 1–7, IEEE, Singapore, at the Marina Bay Sands Expo and Convention Centre, from December 4–8, 2017.
24. Gupta, N., Agarwal, R., Dari, S. S., Malik, S., Bhatt, R., Dhabliya, D. DDoS and cyber attacks detection and mitigation in SDN: A comprehensive research of moving target defense systems. In *2023 International Conference on Data Science and Network Security (ICDSNS)*, pp. 1–8, IEEE, Tiptur, Karnataka, 2023.

25. Nanda, S., Zafari, F., DeCusatis, C., Wedaa, E., Yang, B. Predicting network attack patterns in SDN using machine learning approach. In *Proceedings of the 2016 IEEE Conference on Network Function Virtualization and Software Defined Networks (NFV-SDN)*, Palo Alto, CA, USA, 7–10 November 2016.
26. Wang, J., Wang, L., Wang, R. A method of DDoS attack detection and mitigation for the comprehensive coordinated protection of SDN controllers. *Entropy*, 2023, 25(8), 1210.
27. Ashraf, E., Areed, N. F. F., Salem, H., Abdelhay, E. H., Farouk, A. "FIDChain: Federated intrusion detection system for blockchain-enabled IoT healthcare applications. *Healthcare*, 2022, 10(6), 1110.

# Index

Printed in the United States
by Baker & Taylor Publisher Services